Boiler Shell Weld Repair

Organizing Committee

B J Darlaston
Strutech Consultancy

P E J Flewitt
Magnox Electric

E G Taylor
Magnox Electric

Sponsored by

Co-sponsored by
BNES
INucE

IMechE
Seminar Publication

I MECH E

Boiler Shell Weld Repair
Sizewell 'A' Nuclear Power Station

25 October 1999
IMechE HQ, London, UK

Organized by
The Pressure Systems Group of the
Institution of Mechanical Engineers (IMechE)

IMechE Seminar Publication 1999–14

**Professional
Engineering
Publishing**

Published by Professional Engineering Publishing Limited for The Institution of
Mechanical Engineers, Bury St Edmunds and London, UK.

First Published 1999

ISSN 1357–9193
ISBN 1 86058 244 3

A CIP catalogue record for this book is available from the British Library.

Printed and bound in Great Britain by Antony Rowe Limited, Chippenham, Wiltshire, UK.

Front cover artwork reproduced with the kind permission of BNFL Magnox Generation.

Related Titles of Interest

Title	Editor/Author	ISBN
IMechE Engineers' Data Book	C Matthews	1 86058 175 7
Handbook of Mechanical Works Inspection – A Guide to Effective Practice	C Matthews	1 86058 047 5
A Practical Guide to Engineering Failure Investigation	C Matthews	1 86058 086 6
Integrity of High Temperature Welds	IMechE Conference Transactions	1 86058149 8
Assuring it's Safe: Integrating Structural Integrity, Inspection, and Monitoring into Safety and Risk Assessment	IMechE Conference Transactions	1 86058 147 1
Nuclear Decommissioning '98	IMechE Conference Transactions	1 86058 151 X
Forging and Related Technologies	IMechE Conference Transactions	1 86058 144 7

For the full range of titles published by Professional Engineering Publishing contact:

Sales Department
Professional Engineering Publishing Limited
Northgate Avenue
Bury St Edmunds
Suffolk
IP32 6BW
UK

Tel: +44 (0)1284 724384
Fax: +44 (0)1284 718692

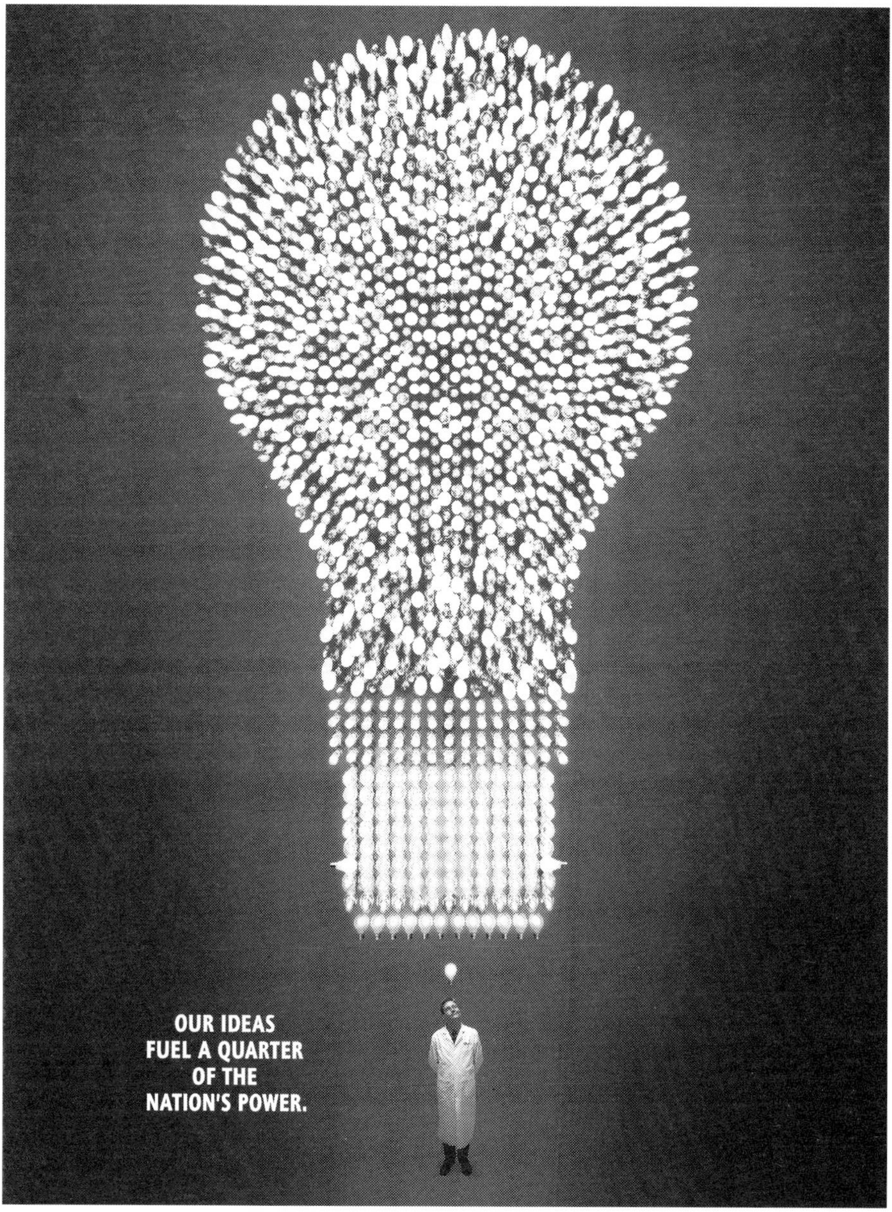

**OUR IDEAS
FUEL A QUARTER
OF THE
NATION'S POWER.**

Over a quarter of the UK's electricity is generated with the nuclear fuel supplied by BNFL.

This helps reduce the country's reliance on fossil fuels - thus saving millions of tonnes of greenhouse gases from being pumped into the atmosphere every year. As these gases are one of the primary causes of global warming, everyone benefits from our contribution to the national grid.

We've also developed technology that recovers around 97% of used nuclear fuel. This can be made into fresh fuel, to burn again.

At the same time, our technologies are helping to decommission the world's outdated nuclear sites in both the East and the West.

What's more, some of our big ideas go beyond the nuclear power industry.

We've created bearings that run silently on a magnetic cushion, which means they never wear out.

We've even produced a liquid that can be breathed, so it can save the lives of premature babies.

A big idea can change the world. And you'll find a world of them at BNFL. To learn more about what we do and how we do it, visit www.bnfl.com or come to one of our Visitors Centres.

BNFL
Where science never sleeps

Contents

Foreword

Over the years, the Pressure Systems Group of the IMechE has provided a programme of seminars and conferences covering many aspects of pressure equipment. Increasing attention is being given to the integration of the elements of technology contributing to the integrity and safety of equipment. The repairs carried out at Sizewell A provide a good example of this integration and the subject complements the other topics of the current Pressure Systems Group programme. Although the boiler in question is part of a nuclear plant the procedures and the approach are equally applicable to non nuclear components.

The Sizewell A boiler weld repairs represent one of the major repairs of pressure equipment completed in the UK in recent years. Although this was undertaken on nuclear components which are ~ 30 years old, modern standards appropriate to nuclear pressure circuits had to be satisfied. To meet these demands the overall project brought together, and implemented, the technological developments that have occurred over a significant period of time. The operators of the plant, BNFL Magnox Generation, agreed to share, with the broader engineering community, their experience of the challenges presented as a result of the multi-discipline approach. This brought together non-destructive inspection, defect diagnosis, structural integrity assessment and in particular the welding technology to secure repairs of the highest demonstrated quality. Moreover these repairs were underwritten against current safety standards applied to nuclear plant in the UK. These standards have to satisfy modern codes, safety case arguments and licensing requirements of the Nuclear Installations Inspectorate. The challenges provide an illustration of the capability of the engineering industry in the UK. The ability to manage large projects, both to meet the technical requirements and also to ensure the commercial viability was demonstrated. In putting together the programme for this one day meeting, members of the project team have been invited to contribute, in particular Mitsui Babcock Energy Limited who worked in partnership with BNFL Magnox Generation to achieve this successful weld repair.

John Darlaston MBE
Chairman, Pressure Circuit Systems Group

S690/001/99

An overview of the need to repair the three boilers

C SMITTON
BNFL Magnox Generation, Berkeley, UK
C J MARCHESE
BNFL Magnox Generation, Leiston, UK

ABSTRACT

When Nuclear Electric Plc set out to justify operation of it's Magnox reactors beyond 30 years in the early 1990's, the Periodic Safety Review (PSR) process, agreed with the NII had a strong focus in engineering assessments to modern standards. For Sizewell A this meant all aspects of plant safety had to be reviewed to these standards and the resulting necessary modifications completed by 1996. One of the major items of the process was a review of the boiler capability to operate for at least the 10 years of the PSR licence extension.

The resulting inspection programme identified the need for repair and it was Magnox Electric Plc, the owner operators at the time, who had to take the decisions leading to the successful repair. The paper will describe the background to the PSR process for justifying safe operation of boilers and the investment for continued operation of the reactor and plant connected to these boilers.

1. INTRODUCTION

Sizewell A, commissioned in 1966, is one of eight power stations now operated by BNFL Magnox Generation and has two Magnox reactor units, cooled by carbon dioxide. The primary circuit, Figure 1, consists of the reactor pressure vessel and four boiler units for reach reactor providing steam to a single turbo alternator. The combined electrical output of the two units is 420 MW net of all on site loads.

The steam generating boiler units shown in Figure 2 were constructed partly off-site and partly on the construction site in the early 1960s from a Ducol W30 type steel. The details are presented in Paper 2. The key construction issues for the boilers were the fabrication from annular rings and endcaps, which were welded from plates and generally furnace heat treated. The plain annular rings sections or courses were first built into larger components and transported to the construction site for final assembly and erection by the Goliath crane on site. Some of the circumferential welds were welded and stress relieved on site, using bands of heater elements. This construction sequence became relevant when it was necessary to review the inspection results some thirty years later.

Site Licence Condition 15 issued by the Health and Safety Executive requires the licensee of a nuclear power station to "make and implement arrangements for the periodic and systematic review and assessment of safety cases" i.e. a Periodic Safety Review or PSR.

To comply with this condition licensees continually review the basis of their safety cases which underpin the safe operation of plant. These reviews cover changes in methods and knowledge, operational experience and also at intervals reviews when the plant is shutdown to undertake specific maintenance and inspection. In addition our arrangements in BNFL require a comprehensive review to be carried out every ten years.

For these reasons as each of BNFL Magnox Generation's Magnox reactors has approached its 30th year of operation, it has carried out a full PSR to gain agreement from the NII to operate for a further ten years. It is worth noting that seven stations have already achieved this and two, Calder and Chapelcross have also carried out PSRs at 40 years to permit 50 year operation.

2.   PERIODIC SAFETY REVIEWS

The content of a PSR is now well established and conforms to models set out by the IAEA (International Atomic Energy Agency). The key elements of a ten year PSR can be distilled down to looking at the current state of the plant and then demonstrating that it is safe to operate for the ten year period. Naturally plant ageing from construction to the current time and how this might continue for the ten year period is crucial. The process brings together historical records of the maintenance, inspection and testing of the plant as well as specific additional requirements in these areas to enable assessments of the condition of the plant in ten years to be made, which are presented as a safety case for further operation.

Of particular importance to PSRs, is a requirement to incorporate modern standards, methods and practices when carrying out these assessments. What was demonstrated either by testing, inspection or calculation at the time of construction as safe, is not necessarily the best practicable that a licensee can or should do towards such reviews 30 years after construction.

3.   THE ROLE OF THE PRIMARY CIRCUIT INSPECTIONS

Each nuclear power station has four systems fundamental to protecting the fission product inventory in the nuclear fuel. They are the fuel cladding material, the pressure circuit, the system to shutdown the nuclear reaction and finally the system that cools and removes the residual heat of the nuclear reaction. In the magnox reactor system the first two are passive and the others active.

Of the passive systems, the fuel is regularly changed so the fuel cladding is replenished. That leaves the pressure circuit as the major system that is passive and not easily replenished.

The boilers or steam generators of the steel reactor pressure vessel Magnox stations are located away from the reactors and outside the biological shield. However, they operate at reactor pressure by containing the primary CO_2 coolant and are connected to the reactor pressure vessel by ducting. The boiler and ductwork have been the subject of regular inspection by NDT to monitor their state since construction. The level of inspection has increased over the years both in the extent of the NDT and the type as technology has increased.

The primary objective of the inspections was to support structural integrity assessments that had shown that even with an assumed level of defectiveness, the structures could withstand fault over pressures well above that of normal operation. Supporting this were the construction records of the quality of construction, over-pressure testing prior to operation and materials testing of the component steels. Usually the materials testing can utilise the experience from similar materials used on other components. For the steels of the Magnox reactor pressure vessels, ductwork and boiler there is a great wealth of information from the 26 reactors built by the UK consortia in the 1950s and 1960s. However there was one exception at Sizewell A, the last of the steel vessel Magnox stations to be built and commissioned in 1966. The exception was the choice of a high strength, low alloy steel used to fabricate the whole of the boiler shell with only some smaller components being made of this steel at other stations.

This uniqueness called for particular attention in the 30 year PSR at Sizewell A. One of the boilers originally constructed, now called boiler 2C, had originally failed a pre-operational hydraulic pressure test, was rebuilt and subsequently passed the same test. The inspection programme started at the beginning of 1996 and would cover a sample, around 20%, of the fabrication welds on each of the boilers starting with those on Unit 1 and continuing with Unit 2. The expectation was that some manufacturing defects would be found but they would not be structurally significant. It was during these inspections that several unusually large defect indications were found on some circumferential welds between the annular rings of three boilers of Unit 2. These observations led eventually to a decision to inspect all fabrication welds on the exterior of the boilers of units 1 and 2, eight in all, as far as was practically possible.

Indeed just over 99% of weld was ultrasonically inspected with the remainder being justified as satisfactory with reference to integrity assessments. These were small lengths where ultrasonic probe access was not possible but in many cases MPI supported the assessments made.

Except for some small welding defects all the boilers of Unit 1 had satisfactory integrity and were unaffected by the type of significant boiler defects found on the boilers of Unit 2. This enabled a satisfactory statement to be made concerning the PSR safety case for Unit 1 and it returned to service in September 1996.

4.   THE NEED FOR REPAIR

The defective areas on Unit 2 were found on three out of the four boilers, with boiler 2B free from such defects able to be used throughout the eventual repair period as a duty boiler for shutdown cooling. The other boilers, 2A, 2C and 2D each had a defect in the same location, between the annular rings numbers 6 and 7, this is referred to as the 6/7 weld. On boiler 2C a further smaller defect was also found on the 5/6 weld. The nature of the defects were evident by MPI as surface breaking cracking much of which was intermittent in nature extending through in some places up to around 40 mm but not penetrating the shell of 57 mm thickness.

As will be discussed in other papers, investigation work, including sampling of the material, conclusively showed that the cracking was due to stress relief cracking during the Post Weld Heat Treatment (PWHT) applied at the time of the shell fabrication stage in the 1960s. Reviews of the records of construction revealed the weld construction details which were subject to radiographic inspections. The code used at the time, BS 1500, required only a pressure test after the PWHT with no further inspection. It is for this reason that such large defects were not found until the advent of extensive ultrasonic NDT.

Extensive work showed that during the 30 years of operation since construction no extension of the defects had occurred. Also the safety of the shells during this period was understood from structural integrity assessments and the proof pressure test applied to the shells before being placed into service. The proof pressure test applied to each boiler at over 1.5 times the current working pressure, was designed to ensure that even if unrevealed defects existed in the structures, survival from the test would demonstrate that the boiler shells would be safe at normal operating pressures and at any fault elevation of these pressures limited by the three safety relief valves on each boiler. Additionally even with the defects found structural integrity assessments showed a significant pressure margin of safety over the current worst fault over-pressure and assumptions on reductions over life of the material properties of the boiler steel.

Nevertheless there was a need to address these defects to demonstrate the highest level of safety if the boilers were to be returned to service. Indeed the objective for operational strategy was to justify ten further years of operation. Several proposals were considered, namely a restraint system around the welds, or a non-PWHT repair, but overall the conclusion was that a full PWHT welded repair, leaving as much of the undefective material as possible, would be used where this was justified. This choice was used for the 6/7 welds but the 5/6 weld defect was amenable to a profiled surface excavation.

5.   THE ECONOMIC NATURE OF THE REPAIR

Repairs of the magnitude presented by the defects found in the Unit 2 boilers presented specific challenges. Although some repairs had previously been carried out on Sizewell A gas ducting involving the replacement of sections and rewelding them into the primary circuit, this repair would be the longest length of a weld repair to an existing and aged structure that we had attempted.

As will be seen in later papers heat treatments in the region of 675°C. to 680°C. on an installed and commissioned boiler would have a significant effect on all the internal and external apparatus. The boiler tube system, the steam drums and thermal shield connections for pipework, could all see higher than normal operating temperatures, which are around 350°C, during the heat treatment. In addition the top duct work connecting the top of each boiler to the reactor pressure vessel was likely to move differentially from the reactor. To prevent adverse effects on the reactor pressure vessel, these top duct sections would be separated from the reactor nozzle sections.

The details of these engineering considerations are dealt with by later papers. However, the magnitude of the site work to implement the repair necessitated a major project framework involving a substantial contracting effort for Magnox Electric plc (latterly BNFL Magnox

Generation Business Group). The external contractor had to have a technical and engineering capability with extensive experience in the material Ducol W30. Much work would be sub-contracted, but a close working relationship would be necessary to both design the repair and develop its implementation in a short time.

The quality of such work is commercially important. However an additional factor would drive the project – it became imperative that the weld repair had to be "right first time". Material of this age that had already passed through a PWHT cycle could possibly not survive many further cycles. It was because of this that a full circumferential repair of the 6/7 welds was attempted, to remove cavitated material as well as the primary surface breaking defects.

The economic considerations dictated that as much parallel working as possible should be attempted to reduce the project length and return the plant to full operation. However as will be seen it became important to feed as much experience gained in one weld repair to another so that complete paralleling of repairs was not possible. In addition, it was logistically difficult to mount simultaneously certain operations on boilers 2C and 2D as they are located in the same boiler house.

Besides the repair, the PWHT presented a significant challenge in itself. The high temperatures needed to stress relieve the weld repairs had to be deployed to the boiler shells in situ and controlled to a much higher degree than in the original construction to avoid similar stress relief cracking again. Overall the engineering and technical work had to be supported by a quality assurance system involving a significant element of peer review and third party inspection. All was justified against the successful outcome which economically would secure at least ten more years of output from Unit 2 at Sizewell A.

6. THE ROLE OF THE REGULATOR

The engineering and safety considerations necessary for the return to service of Unit 2 at Sizewell A included a satisfactory PSR. The NII as the regulator have a responsibility to ensure that the reviews meet modern standards and for the Unit 2 boilers provided guidance on the acceptability of the approach to the repair. As will be seen their involvement extended from knowledge of the initial inspections, through the design and implementation of the repairs and the recommissioning of the plant.

Although the NII is independent, the joint need of BNFL as customer and NII as regulator, to have acceptable Third Party Inspection besides independent verification on safety assessments led to a licensing structure that gave BNFL staged agreement to proceed throughout the project. As will be seen from other papers the planning of assessment schedules to permit adequate time for regulatory assessment would turn out to be as important as the planning of other parts of the project.

7. SUMMARY

After rigorous inspection of all main boilers at Sizewell A, three Unit 2 boilers were found to have significant defects that could most reliably be addressed by a PWHT welded repair. Although operation for 30 years with these defects was justifiably safe, the desirability of further operation to a high degree of confidence led to the decision to weld repair three of the defects and profile a fourth.

The repair would be a major undertaking both in the required technical quality and verification.   It would require a major logistics undertaking at the site and a major contracting exercise was necessary to match the exacting standards to be met.

To achieve an acceptable agreement from the Regulator the level of independent verification was a critical factor.   The economic driver for return to service of Sizewell A Unit 2 also meant that the repairs had to be completed successfully to a high standard first time.

8.   ACKNOWLEDGEMENT

This paper is published with the permission of the Director Technology and Central Engineering, BNFL Magnox Generation.

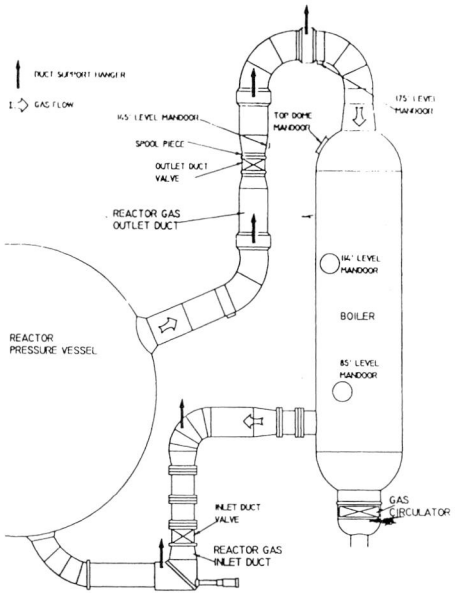

Figure 1: Schematic layout of gas circuits at Sizewell A

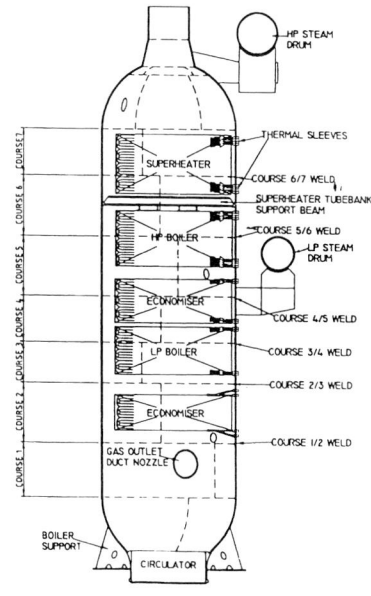

Figure 2: Schematic layout of boiler at Sizewell A

An evaluation of the nature and origin of cracking and implications for the repair strategy

L F EXWORTHY, B J C ELLIS, and **P E J FLEWITT**
BNFL Magnox Generation, Berkeley, UK

ABSTRACT

Following a period of satisfactory service the seam welds in the low alloy ferritic steel boiler shells of reactor 2 at Sizewell A Power Station were, as part of the Periodic Safety Review requirement, subject to a non-destructive inspection during 1996. These revealed defects in the circumferential welds and the subsequent comprehensive ultrasonic inspections of all boilers provided the size and distribution of the cracks. A range of investigations, including microstructural, chemical composition, hardness and fractography, were undertaken, both at site and on excavated samples to diagnose the factors leading to cracking. The results are discussed with respect to the mechanism of cracking and implications for the repair strategy.

1. INTRODUCTION

The boiler shells at Sizewell A Power Station; Fig. 1, were fabricated by Babcock and Wilcock in the period 1961 to 1963 from 57 mm thick ferritic steel plate supplied to their own specification (1). This is a semi-air hardening, low carbon alloy steel, BW87A, with a nominal composition Fe – 0.1C – 1.4Mn – 0.5Cr – 0.25Mo – 0.2Ni – 0.1V, equivalent to Ducol W30 but of a lower carbon concentration. This steel can be normalised and tempered. It was selected to have superior mechanical properties to the C-Mn steels used for the rest of the tranche of Magnox power stations to enable the boilers at Sizewell to be constructed from thinner section plate. Each boiler is an 18.9m long cylinder, of internal diameter 6.86m, to which hemispherical dome ends are fitted. The body was fabricated from seven cylindrical courses each made from three plates joined by axial welds and these courses were then circumferentially welded together. All these butt welds were of double V-preparation geometry. The locations of the axial welds were alternately staggered by about 100 mm to avoid a continuous axial seam in the boiler. The vertically aligned plates were arbitrarily designated as A, B or C and the seam welds diametrically opposite each were identified by course number and plate designation, i.e. axial weld 7B joins plates A and C in course 7. Six equally sized petal plates were welded together to form the bottom dome whereas the top dome was fabricated from two rings of petal plates, six in the lower rings and four in the upper.

All axial welds were made at works using a submerged arc welding (SAW) process and Unionmelt 4 filler wire. The individual courses were furnace heat treated for 3 hours at 600 °C to stress relieve the axial welds and temper the plates which had been supplied in the normalised condition. Circumferential welds 3/4 and 6/7, Fig. 1, were also made at works by manual metal arc welding (MMA), using Babcock and Wilcock H1 electrodes and stress relieved. It is not clear from the archive records if the heat treatment was undertaken in a furnace, either at the same time as the axial welds or subsequently, or by localised heating. The courses were transferred to site where the remaining circumferential welds were completed, again by MMA welding and locally stress relieved by electrical induction coil heating with supplementary electrical radiant heating (1). The completed welds were examined by radiography prior to post weld heat treatment and repairs made accordingly. Finally, a proof test was undertaken at 1.8 times operating pressure.

Boiler 2C failed during the proof test by the propagation of brittle cracks through the B plates on courses 4 to 7. The failure was attributed variously to shock loading arising from the collapse of a support chock (1) or a combination of a pre-existent welding defect in a thermal sleeve weld, poor plate properties and localised bending stresses imposed by the boiler supports (2). Following the proof test failure, courses 4 to 7 of the boiler shell were sub-divided by gas cutting through the centre-line of the axial and circumferential welds and returned to works. The four cracked B plates were discarded and replacements obtained. The four individual courses were rebuilt using the original A and C plates. The radiographic records state that SAW was employed to remake the axial welds. Courses 6 and 7 were re-welded at works by MMA. The stress relief heat treatment was undertaken at works, possibly at a temperature of 650°C rather than 600 °C. The remainder of the circumferential welding to rebuild the boiler was carried out at site by MMA and the welds were stress relieved at 600 °C by local induction heating; the inspection requirements were as the original construction.

Following entry into service in 1966, the boilers performed satisfactorily at a maximum temperature, at weld 6/7, of ~ 375 °C. However, in 1996, as part of the Periodic Safety Review requirement, the top dome welds and the vertical seam welds in courses 6 and 7 of all boilers were subjected to non-destructive inspection. During magnetic particle inspection of the external surface of the vertical welds 6B and 7B of boiler 2C, three linear indications were detected on the course 7 side fusion boundary of the circumferential weld between courses 6 and 7 (3). These defects were of lengths 15, 13 and 3 mm within a circumferential length of 83 mm, and subsequent ultrasonic inspection revealed a sub-surface defect 25 mm in depth and over 4m in circumferential length. The discovery of these defects eventually led to the non-destructive inspection of all the seam welds in boiler 2C and all other boilers at Sizewell A Power Station.

In this paper we summarise the results of comprehensive non-destructive inspections, Section 2, undertaken to provide the size and distribution of cracks identified within the boiler shells. Detailed microstructural, chemical, hardness and fractographic investigations undertaken, both at site and on excavated samples, to diagnose the underlying mechanisms leading to the observed cracking are described in Section 4. The results are discussed in Section 5 with respect to the mechanism of cracking and the implications for a repair strategy.

2. NON-DESTRUCTIVE INSPECTION

A comprehensive magnetic particle and manual ultrasonic inspection of all the circumferential welds in all boilers was completed following detection of defective areas on boiler 2C. However, when defective areas were identified more detailed inspections were undertaken by automatic scanning using jigged probes and the MIPS/Micropulse ultrasonic data acquisition system and GUIDE processing system (4)(5). The magnetic particle and original manual ultrasonic inspections had reporting thresholds of > 20mm linear length, unless intermittently longer or with measurable sub-surface extent ≥ 3mm. The reporting threshold was reduced for the automated ultrasonic inspection. These inspections revealed defects in boilers 2A and 2D in addition to those in 2C and the results for the circumferential welds are summarised in Table 1. All of the sub-surface defects were estimated to lie parallel and close to the weld fusion boundary as projected from drawings of the original weld preparations. With the exception of one small sub-surface defect, all of the defects in boiler 2C were confined to the C plates. On the course 6 side of the 6/7 and 5/6 welds the defects lay within the circumferential extent of the defect on the course 7 side of weld 6/7. On boilers 2A and 2D the defects were confined to the A plates with coincident cracking on both fusion boundaries on boiler 2D.

3. DIAGNOSTIC TECHNIQUES

To diagnose cracking associated with weldments in the BW87A ferritic steel boiler shells a variety of techniques were used, both at site and for samples extracted from the boilers.

3.1 Metallography

At site after preparing the region of the weldment surface to be investigated by normal metallographic procedures the surface was etched with 2% Nital reagent to reveal the required microstructural detail. To ensure that fine detail such as cavitation was captured a multiple polish/etch technique was employed. The surface was then overlaid with a cellulose acetate film to replicate the microstructural detail (6). Discs (3mm dia.) of the parent plate and weld metal removed from extracted samples were thinned for transmission electron microscopy by jetting in a 5% perchloric acid and 95% ethanol solution maintained at room temperature with an applied voltage of 18V and a current density of $\sim 10^4 \text{Am}^{-2}$. The foils were examined at 200KeV using a JEOL 3010 transmission electron microscope and in a Vacuum Generators VG 501 FEG – STEM fitted with an energy dispensive spectrometer. To analyse the chemical composition of grain boundaries samples were fractured at $-196\ ^\circ$C and examined in a Fisons Minilab – 310F Auger electron spectrometer (AES). Fracture surfaces were examined in a JEOL 840A scanning electron microscope operating at 15KeV in the secondary electron imaging mode. When chemical analysis was not possible the composition of samples was established using a JEOL 8600 Super Probe electron microprobe analyser (EPMA) fitted with wave length dispersive crystal spectrometers and operating at 30KeV. The chemical composition was evaluated using a ZAF correction program (6).

Metallographically prepared specimens were examined optically to measure oxide thickness at discrete positions along cracks to establish the mean thickness. Oxide thickness and exposure times for oxidation in air for a range of ferritic steels can be related by a parabolic rate equation for temperatures up to ~600 $^\circ$C (7), where exposure time, t, is given by:

$$t \quad = \quad (x^2/P) \tag{1}$$

where P is the rate constant for the particular steel and x is the oxide thickness.

3.2 Sample Extraction

Two types of sample were extracted from defined locations within the boiler shells, scoops and boats. The scoops were extracted using a hemispherical cutter to give discs ~25 mm dia. which extend to a maximum depth of ~4 mm. By contrast boat samples of varying length and depth up to ~25 mm allowed material containing critical features of the weldments and associated cracking to be extracted. A detailed description of the boat sampling method is given elsewhere (8), but it is based upon a cutter made from a thin saucer-shaped abrasive disc which rotates at high speed. Two cuts intersect sub-surface along the length of the boat sample which has axes coincident with the boiler circumference and the mid-length position of maximum depth sited over the feature to be extracted. A total of eight boat samples were extracted at locations based on the results of the automated ultrasonic inspections.

3.3 Hardness Testing

Hardness measurements were undertaken on extracted samples using a Vickers macro-hardness tester and a Mitutoyo MVK-H1 micro-hardness tester. In-situ measurements were made on the boiler shells with an "Equotip" portable hardness tester and the results converted to Vickers diamond pyramid values.

4.   DIAGNOSTIC INVESTIGATION

4.1 Introduction

Investigations were undertaken in two stages, subsequent to non-destructive inspection of the weldments, to diagnose the nature and mechanism of the cracking. The first stage covered on-site investigations and limited sample removal and this was followed by a more comprehensive programme of sample removal. The former was directed to providing a preliminary diagnosis of the microstructural condition of the weldments and

4.2 Stage 1 Investigation

4.2.1 Metallographic Replication

On boiler 2C substantial defects were ultrasonically detected at both fusion boundaries of weld 6/7 and on the course 6 side of weld 5/6. In all three instances the buried defect intermittently broke through to the external surface. Visually the defects were similar, lying parallel to and on, or very close to, the fusion boundaries and frequently exhibiting significant gape. Weld repairs to the plate edges and/or the remnants of attachment welds were often observed. Examples of cracking on all three affected fusion boundaries were metallographic replicated and the results are summarised below.

Defects lay either in the coarse grained HAZ or in the weld metal close to the fusion boundary and frequently crossed from one microstructure to the other. At the crack tips the path was invariably intergranular and often associated with extensive grain boundary cavitation, the latter always being more prolific in the coarse grained HAZ. When cracks were within the weld metal, micro-cracking and cavitation were commonly present in the adjacent HAZ. Remote from the tips the crack increased in gape and the path became less obviously intergranular and often indeterminate. Occasionally, however, the main crack was accompanied by subsidiary, intergranular cracking and cavitation both in the weld metal but more usually in the coarse grained HAZ. Toward the tips the cracks were fully occluded with oxide. Where the cracking exhibited significant gape the crack surfaces were heavily oxidised but occlusion had not occurred. Localised weld repairs/attachment weld remnants and arc strikes were observed both adjacent to, and abutting the main weld. The cracks were occasionally diverted into their HAZ's but their orientation was still primarily circumferential. The defects on boilers 2A and 2D had very similar features. In summary, metallographic replication the defects on all three boilers revealed the cracks formed or extended by interlinkage of discrete grain boundary cavities consistent with a creep mechanism. Since the boilers operate at a temperature of $\leq 400\ ^\circ C$, in service creep leading to cavitation is unlikely and this damage is consistent with stress relief cracking during post weld heat treatment being the responsible mechanism.

Replication of the parent microstructure of each of the six plates, comprising courses 6 and 7 on boiler 2C, revealed that all were tempered bainite colonies within a ferrite matrix with a grain size of ~30μm (mean linear intercept); there were no significant differences between the plates. The volume fraction of bainite was typically ~40% of the total microstructure. Replication of parent plates from other boilers, including 2A and 2D, confirmed that this microstructure was typical of all plates, both cracked and uncracked.

4.2.2 Metallography of extracted samples

During Stage 1 there was a need to maximise the information gained by taking the largest samples possible from boiler 2C commensurate with removing the minimum of sound material. As a consequence, the maximum depth of excavation was limited to ~15mm and potential sampling sites restricted to those where the sub-surface depth of defects was

cracking, whereas the latter focused on factors influencing cracking.

considerably in excess of this value. Ideally samples would have been taken from all three fusion boundaries which exhibited cracking but, for example, in the case of weld 5/6, Table 1, the need to accommodate the above criteria would have resulted in a small sample of limited value. In the event, a 150mm long boat shaped sample of 13 mm maximum depth was extracted from each fusion boundary of weld 6/7, boiler 2C.

The extent of cracking on the extracted sample from the 6 side fusion boundary, as revealed by MPI, is shown in Fig. 2. Sub-surface the major cracking was almost exclusively in the coarse grained HAZ immediately adjacent to the weld fusion boundary, Fig. 3. It passes through weld metal following the shortest route to the external surface by breaking through the wide capping run on the weld toe. This feature doubtless accounts for the switching of the crack path between HAZ and weld metal observed in surface examinations. The cracking in the HAZ was clearly intergranular and mainly confined to the coarse grained regions, Fig. 4(a). Subsidiary cracking and cavitation, in varying amounts, was observed adjacent to the crack over the full length, Fig. 4(b). The most prolific cavitation was encountered at the near surface where it was within weld metal and the deepest part of the crack where it terminated in fine grained HAZ. Cavitation and incipient intergranular cracks were often associated with the weld metal particularly where the main crack broke through the toe run weld bead to the free surface.

The sample extracted from the 7 side fusion boundary again revealed cracking in the HAZ close to the fusion boundary, similar to that shown in Fig. 4(b). Although the main crack did not appear obviously intergranular the small subsidiary cracks and cavitation, frequently seen adjacent to the main crack, were clearly so. It is interesting to note that very little coarse grained HAZ was in evidence. However, the grain size inferred from the subsidiary cracking was very similar over the crack depth sampled, irrespective of the HAZ grain size. In the grain refined HAZ the inferred grain size was much larger than the existing grain size and more consistent with that in the coarse grained HAZ. In these instances cracking had clearly preceded the grain refinement.

Generally the cracks have characteristics associated with stress relief cracking occurring during post weld heat treatment. Certainly this is consistent with the observed morphology and location of the cracks, including the fact that cracking was on both fusion boundaries of the same weld. Not all of the subsidiary cracks, however, were accompanied by cavitation. Some were short with a comparatively wide gape and were more typical of hydrogen cracks. Furthermore, on some parts of main cracks there were no subsidiary cracks or cavitation; the latter could not be detected even after many cycles of polish-etching. It is difficult to reconcile this lack of damage with stress relief cracking, hence in some cases hydrogen cracking may act as initiators for subsequent stress relief cracking. Finally, if the prior austenite grain boundary cracking in the fine grained HAZ, immediately adjacent to the fusion boundary, occurred during the stress relief heat treatment when this area was originally coarse grained, then the weld runs responsible for the subsequent grain refinement must have been deposited after heat treatment. Thus, weld repairs were either made after the second stress relief, during the boiler rebuild, or some cracking, possibly minor, in the coarse grained HAZ was generated during the original stress relief heat treatment and was not completely removed when the original weld was cut-out. To examine the factors which may promote stress relief cracking, including grain strengthening and/or grain boundary embrittlement, further microstructural investigations were undertaken in Stage 2.

4.2.3. Oxide Dating

The oxide thickness on the majority of the continuous cracking, in both extracted samples, was generally uniform with depth from the external surface indicating that the crack surfaces had been exposed to an oxidising environment for a similar length of time. On course 6 side the oxide was 35 to 40 µm thick and on the course 7 side 25 to 30µm thick. For the first 17318 hours of the life of boiler 2C, the reactor gas outlet temperature, T_2, varied between 362 °C and 394 °C subsequently reducing to 360 °C for the remaining 171000 hours of its operational campaign. To simplify calculations a weighted average temperature for a T_2 of 390 °C will be assumed pessimistically for the initial operating period. The temperature at weld 6/7 can be estimated as 375 °C and 348°C for a T_2 of 390°C and 394°C respectively. Employing the appropriate parabolic rate constants in equation 1, see Section 3.1, the initial 17318 hours at 375 °C would produce 11.8µm of oxide. To achieve the measured thickness of oxide on course 7 side and course 6 side would take a further 56000 to 107000 and 174000 to 256000 hours of operation respectively at 348 °C. Hence for the course 6 side it would be reasonable to claim that cracking pre-dates boiler commissioning. In the case of the course 7 side, 25 to 30µm of oxide would be expected after 201000 to 290000 hours exposure at 348 °C and thus the crack surfaces were exposed to the oxidising environment very early in the boiler life. However, given the inherent errors in oxide growth rate predictions and the assumed temperatures which take no account of thermal gradients through the boiler shell wall, the measured oxide thickness is consistent with exposure throughout the life of the boiler. The difference in oxide thickness at the two locations could be a consequence of small differences in chemical composition of the two plates.

4.2.4. Fractography

Fractographic examination of the crack surfaces in a scanning electron microscope was largely confounded, as anticipated, by the presence of the relatively thick coating of oxide (9). Chemical stripping of the oxide in inhibited hydrochloric acid confirmed that the crack path was intergranular, but any underlying detail had been obliterated by oxidation. Specimens containing crack tips were fractured in liquid nitrogen. Cavitation damage was most likely to be present in the coarse grained HAZ, but unfortunately most of what little material there was ahead of the crack tips fractured in either weld metal or fine grained HAZ. Nevertheless one small area of cavitation on a grain boundary in coarse grained HAZ was revealed, see Fig. 5.

4.2.5. Hardness Measurements

Hardness measurements made in the columnar grained structure of each weld bead of the samples extracted from the course 7 and 6 sides were similar, varying from 226 to 241 Hv. Values immediately adjacent to the fusion boundary, in both the coarse grained and wholly refined HAZ on the course 7 side, were close to 300 Hv for the former and 270 Hv for the latter. These were markedly higher than the maximum value of 243 Hv recorded in the coarse grained HAZ on the course 6 side of the weld. To ascertain if these differences were significant, comparisons were made with hardness measurements made at site on axial and circumferential welds of the top two courses of the boilers of Reactor 1, Table 2. For the axial welds HAZ values of 159 to 308 were recorded and for the circumferential welds values of 207 to 286. Substantial differences were often found between the two HAZ's on the same weld, although it should be appreciated that the highest hardness regions of each HAZ may not have been accurately targeted; indeed, some lower values are indicative of parent plate. Nevertheless hardness values approaching 300 Hv are typical of stress relieved HAZ. Weld metal hardness was measured and provided values in the range of 236 to 250 Hv, with one exception of 210 Hv.

4.2.6. Chemical Analysis

Chemical analyses were undertaken using an EPMA for the parent plate and weld metal contained within the two samples extracted from weld 6/7. An area 100μm x 100μm was sampled and the results confirmed compositions consistent with the specification for BW87A and Babcock and Wilcox H1 weld metal; no rogue materials were present. For the two plate steels these analyses suggest a slightly higher alloying content than the case history records. This is not surprising as the EPMA incident electron beam samples very small volumes of material (6) and the differences may relate to inherent local composition variations.

4.3. Stage 2 Investigation

4.3.1. Introduction

This Stage 2 investigation considered:

(a) Scoop samples extracted from cracked and uncracked plates to address susceptibility to cracking.

(b) Boat samples, up to 26 mm deep, extracted from all cracked weldments, except 2C 5/6, to provide more detail for crack initiation and growth mechanisms and a measure of the spatial distribution and extent of grain boundary cavitation damage beyond the main crack envelope.

4.3.2. Scoop Samples

The locations of the scoop samples and a summary of the results of optical metallography are given in Table 3. All but the last four of the scoop samples listed in Table 3 were taken from the external surface of the boiler. Optical metallography confirmed the microstructure of the low alloy ferritic steel plates was consistent with the results of replication, Section 4.2, a mixture of bainite colonies and ferrite grains, Fig. 6. For each scoop sample the overall microstructure was characterised by measurement of the ferrite grain size (mean linear intercept), the area fraction and hence volume fraction of bainite, and hardness (Hv). The hardness of the parent plate varied from 185 to 246 Hv and although there was no correlation between hardness and grain size the former did increase with increasing volume fraction of bainite, Table 3. There was also no obvious correlation between either volume fraction of bainite or hardness of the parent plate with occurrence of cracking. Moreover, there was no evidence of any microstructural abnormalities which could have been influential in terms of HAZ cracking susceptibility.

All but one of the scoop samples contained regions of coarse grained HAZ, the exception containing only refined grained HAZ, Fig. 6. There were no obvious features which could uniquely be related to incidence of cracking. Within the HAZ the prior austenite grain was not always clearly delineated so that, rather than measure grain size by linear intercept, an optical comparison with ASTM nominal grain size charts was undertaken. Notwithstanding potential inaccuracies there was no correlation between grain size, hardness and incidence of cracking. The hardness of the HAZ frequently varied substantially within the coarse grained HAZ of a single scoop so maximum hardness values are provided in Table 3, rather than average values. The measurements were made with a calibrated microhardness tester. Maximum hardness values ranged from 285 to 376 Hv and whilst one of the 285 Hv values corresponded to grain refined HAZ, similar low values were recorded for coarse grained HAZ. Significantly, the highest hardness did not correlate with cracking susceptibility. Indeed, for the scoops extracted from the external surface of the boiler, the cracked plates possessed maximum HAZ hardness values ranging from 285 to 317 Hv. On the internal scoops values of 325 and 340 Hv were recorded on cracked plates at positions remote from the cracking.

Transmission electron microscopy of plate specimens provided more detail of the equiaxed ferrite grains ~10 μm dia. and tempered bainite microstructure. In addition to M_3C type carbide precipitates these microstructures contained distributions of plate-like MC type carbide precipitates which were often aligned, Fig. 7(a). Chemical microanalyses showed these to contain both V and Mo in the at % ratio 1:1. The microstructure of the HAZ was bainite with a lath arrangement, Fig. 7(b). The laths were typically between 0.5 and 2 μm wide with M_3C type, where M is Fe and Mn, carbide precipitates on the boundaries. In addition there were plate-like MC type carbide precipitates similar to those in Figure 7(a). Within both the plate and HAZ samples there was no evidence of precipitation denuded regions adjacent to grain boundaries. Brittle fractured samples of plate and HAZ material were examined in the AES to measure the composition of the intergranular fractured regions. Figs. 7(c)(d) show a secondary electron image of the fracture surface and the corresponding phosphorous distribution map. Typically for the plate and HAZ microstructures phosphorous levels of 0.15 to 0.25 of a monolayer were observed. As a consequence there was some evidence of impurity element segregation to ferrite or prior austenite grain boundaries which may promote the formation of creep cavitation after this period of service (10) (11). Other minor elements such as Ni, Cu, Sn, As, Sb were not found.

Six scoop samples removed from plate, Table 3, were subjected to comprehensive chemical analysis which confirmed them to be consistent with specification (12). Although there were small differences they were not systematic and there is no correlation between chemical composition and cracked plates.

4.3.3. Boat Samples
A total of six boat samples were taken from the following locations based on the results of automated ultrasonic inspections: boiler 2A-6/7-7A, boiler 2C-6/7-7C, boiler 2C-6/7-7C, boiler 2C-6/7-6C, boiler 2D-6/7-7A and boiler 2D-6/7-6A. A transverse section at the mid-length position of each boat sample, i.e. deepest point sampled, was prepared for

metallographic examination. This preparation included a multiple polish/etch technique to ensure that the full extent of any creep cavitation damage would be revealed. It was immediately evident that the cracking observed was very similar in nature for all six samples and to that seen in the previous boat samples from 2C-6/7, Section 4.2. Almost invariably cracks lay in the HAZ immediately adjacent to the fusion boundary, although some instances of weld metal cracking were also observed. Whilst the crack path of some of the major cracks was difficult to determine, that of the finer cracks and much of the extensive subsidiary cracking was intergranular and associated with grain boundary cavitation. As the cracking increased in depth it became progressively finer and discontinuous.

An example of the distribution of cracking and cavitation on grain boundaries is shown schematically in Fig. 8. Typically, larger cracks traversed both coarse grained and grain refined HAZ. However, these eventually terminated in grain refined HAZ and the subsequent discontinuous 'pockets' of cracking and cavitation were confined to coarse grained regions. As cracking became discontinuous the severity rapidly reduced within each successive pocket, i.e. with increasing depth, until only grain boundary cavities and perhaps a number of discrete micro-cracks of a few grain diameters in length were observed. Unfortunately, because of the need to limit the sampling depth to secure a future repair, none of the samples contained a further pocket of undamaged coarse grained HAZ. Nevertheless, the cessation of damage, whilst still in relatively coarse grained HAZ, adds credence to the inference that the presence of cavitation alone indicates that there will be very little, if any, damage at greater depths. From a comparison of all samples it is evident that the most severe cracking was located in the coarse grained HAZ of the weld run immediately beneath the capping run. Although the majority of the subsidiary cracks were intergranular a few had a path that was either not obvious or transgranular. On some parts of the major cracking, where the path could not be determined with confidence, the degree of associated subsidiary cracking was much reduced. Both of these features were observed in the Stage 1 boat samples from 2C – 6/7 and are consistent with hydrogen cracking. Whilst limited hydrogen cracking may be present it is less extensive than initially deduced from the Stage 1 examination.

Hardness measurements in the coarse grained HAZ were hampered by the presence of extensive subsidiary cracking but, nevertheless, by employing microhardness testing (100gm load), maximum hardness values in this region of the HAZ of 243 to 332 Hv were recorded. No significant differences in hardness with increasing depth beneath the surface were found.

During the examination of the Stage 1 boat samples from 2C – 6/7 oxide thicknesses of 25 to 30µm and 35 to 40µm were measured on the 7 and 6 side cracks respectively consistent with the cracks having originated during boiler fabrication. Oxide thickness measurements on the current samples were, boiler 2A-6/7-7A, 25µm, boiler 2C-6/7-7C, 30µm, boiler 2C-6/7-6C, 75µm, and boiler 2D-6/7-6A, 37µm. Less oxide was observed on the remaining samples, but since the defects were not surface breaking access of air would have been restricted. These thickness values are equal to or exceed those measured previously, see Section 4.2.3, and support the view that cracks pre-date boiler operation.

5.   DISCUSSION AND CONCLUDING COMMENT

5.1. Nature and origin of cracking

The cracking within the BW87A circumferential weldments is primarily in the coarse grained regions of the HAZ, although cracks in two such regions frequently link through the intervening grain refined HAZ. However, the most susceptible location for cracking is in the coarse grained HAZ of the first weld bead immediately beneath the capping run. In all instances the cracking becomes discontinuous with increasing depth and contained within isolated pockets of coarse grained HAZ. Furthermore, once cracking has become discontinuous the extent of damage in successive pockets reduces rapidly with increasing depth. It is encouraging to note that in most cases the upper bound defect depth predicted by the ultrasonic inspections encompassed both the major cracking and the discontinuous damage observed in the diagnostic investigations. Stress relief cracking occurs in the HAZ of weldments during the post weld heat treatment cycle (13). It is a grain boundary failure phenomenon that arises from the relief of the welding induced residual stresses. The residual stresses are relaxed by strain accumulation within the body of the grains, and usually results in grain boundary sliding. However, if the material cannot relieve these stresses by strain accumulation then grain boundary damage can intervene leading to intergranular fracture. Since grain boundary sliding is more readily accommodated in fine grain size material, coarse grain regions of the HAZ are more susceptible to intergranular failure, as shown by this investigation. Overall susceptibility to stress relief cracking will be controlled by the conjoint action of the stresses, the material microstructure and properties and the heat treatment cycle as summarised in the Venn diagram, Fig. 9(a). Where these three contributions interact then stress relief cracking occurs; removal or modification of one of these contributions eliminates or reduces the likelihood of cracking.

In weldments that are not susceptible to stress relief cracking during the post weld heat treatment cycle, the strain accumulation by a creep deformation process is accommodated between grains by a combination of grain boundary sliding, vacancy diffusion and dislocation climb. If, however, cracking occurs, damage initiates at discontinuities in the grain boundaries such as ledges or at grain corners producing either discrete grain boundary cavitation or, in certain circumstances, growth over a complete boundary leading to decohesion. If there is a weaker precipitate denuded region adjacent to the grain boundaries nucleation and growth of cavities can be promoted. However, depletion was not observed in the present weldment HAZ regions. There are a large number of detailed mechanisms that have been derived to describe nucleation and cavity growth, but at the stresses and temperatures encountered during a stress relief heat treatment, plasticity controlled growth is more likely. The basic assumption in this model, referred to as power law growth, is that the cavity growth rate is given by an equation of the form (14):

$$(dr/dt)_p = \quad r\dot{Y} \quad - (\tilde{o}/2\mu) \quad (2)$$

where $\dot{Y}$ is the strain rate, μ is the coefficient of viscosity in shear, $\tilde{O}$ is the surface energy and r the cavity radius. Certainly power law growth will be promoted by larger cavity radii, high strain rates, and a reduction in $\tilde{O}$. The latter would result from grain boundary impurity segregation which would not be present in these boiler weldments at the post weld heat treatment temperature of $> 600^{0}C$. The rate of change of cavity radius with strain, $°$, may be expressed as:

$$(dr/d°)_p \quad = \quad r \quad - \quad (3\tilde{O}/2\sigma) \qquad (3)$$

where σ is the applied uniaxial stress.

It has been shown to apply providing the nucleated cavity exceeds a critical radius (15). For the local strain accumulated in the present HAZ of thickness ~1mm, the post weld heat treatment cycle which includes a dwell at $600^{o}C$ for ~4 hours and a residual tensile stress of ~400 MPa, the critical radius will be ~1µm; a realistic value for cavity growth. Hence the stress relief cracking follows the sequence of damage accumulation in the HAZ of the weldments, (i) nucleation of discrete cavities on grain boundaries usually oriented normal to the maximum principal stress and at triple boundary junctions, (ii) further nucleation and growth of these individual cavities which link to form micro-cracks over one or more grain boundaries, (iii) interlinkage of many grain boundaries leading to the formation of aligned macro-cracks.

To describe the magnitude and distribution of the stress relief cracking observed in the boiler shell weldments of Reactor 2 further contributions have to be considered. Two stress profiles exist, welding induced residual stresses and stresses arising from the localised post weld heat treatment. These two profiles have the general form with depth relative to the double V butt weld shown in Fig. 9(b). It is the combined tensile stress operating during the heat treatment cycle which limits the depth, from the outer surface, of the cavitational damage and hence cracking. These stresses are a maximum a few mm below the surface where cracking occurred preferentially. Strain accumulation is the critical parameter and since the thermal stresses are longer range than residual stresses they are capable of generating significant strains through elastic follow-up. Hence cracks may be produced within the region of substantial tensile stress which, in this case, extends to the order of a quarter of the wall thickness, Figure 9(b). In summary, stress relief cracks may develop within the HAZ of the BW87A low alloy ferritic steel, particularly within coarse grained regions of the microstructure, leading to cavitated regions of up to ~15mm depth. However, the microstructure of the HAZ is variable with depth and, indeed, there is a periodic change from coarse to fine grain microstructures, Figure 8. This influences the formation of the macrocracks and the distribution of the grain boundary cavitation developed during the stress relief heat treatment. Thus the observed distribution is consistent with the qualitative model proposed by Dolby and Saunders (16).

The examination of the scoop samples failed to reveal any differences in microstructure, hardness or chemical composition of the parent plates which would account for the observed cracking. Susceptibility of low alloy ferritic steels to stress relief cracking, ΔG, can be estimated from their composition in wt % by the empirical relationship (17):

$$\Delta G = Cr + 3.3Mo + 8.1V - 2 \qquad (4)$$

If ΔG is greater than zero a steel is considered to be susceptible; the more positive ΔG the greater the susceptibility. Values of ΔG were calculated for all the plates in courses 1 to 7, inclusive, for all 8 boilers via the case history records of test certificate chemical analyses and found to be in the range 0.07 to 1.0. A comparison of these values with those for the cracked plates is given in Fig. 10(a). It is clear that the cracked plates were by no means exceptional or unduly susceptible. A value of 1.0 for course 2, plate A on boiler 2B is generated from an exceptionally high vanadium content of 0.18% and it is suspected that an original more realistic level of 0.13% may be appropriate; a transfer error. If this value is excluded the maximum value for ΔG is 0.711 for course 4, plate C on boiler 2A. ΔG values of 0.542 and 0.54 for the C plates on courses 6 and 7 respectively of boiler 2C are toward the top of the range but are by no means unique; some 51 plates out of a total of 168 having a value in excess of 0.5. Interestingly, all but 7 of these were on the Reactor 2 boilers. The susceptibility of low alloy ferrite steels to hydrogen cracking, all other welding conditions being equal, can be ranked by an empirical correlation based upon composition, wt %, to derive the carbon equivalent, CE, (18) where:

$$CE = C + Mn/6 + (Cr + Mo + V)/5 + (Ni + Cu)/15 \qquad (6)$$

Values of between 0.47 and 0.57, Figure 10(b) for the boiler plates indicated that all plates, including cracked plates, have low hydrogen cracking susceptibility.

Given that there were no significant metallurgical differences between cracked and uncracked plates and that susceptibility of this low alloy steel to stress relief cracking is not particularly high, the question remains as to why certain plates cracked. Whilst hydrogen cracking may be influential as an initiator this contribution, from the evidence, is minor and other factors relating to the welding operation and the stress relief heat treatment would have been more significant. The observation of cavitational damage in some sub-surface weld beads is relevant since weld metal is less susceptible to stress relief cracking than the parent plate HAZ. One inference from the presence of this damage and the extent of the subsidiary cracking and cavitation in the HAZ is that these regions were subjected to particularly high levels of strain. In addition it has to be recognised that the HAZ of the parent plates was of low creep ductility for the specific post weld heat treatment.

Oxide dating supports the cracks forming during the stress relief heat treatment cycle. Furthermore, all of the crack tips observed terminate in association with grain boundary cavitated material. Certainly creep rupture by cavitation does not occur in BW87A type materials at temperatures as low as 360 °C (0.2Tm) and therefore none of the cracks examined would have been produced during boiler operation. During the early life of the boiler the operating temperature was ~ 390 °C and in the vicinity of the cracks this would have been 375 °C and growth by a creep cavitation remains unlikely. Crack extension by other creep mechanisms can occur at these temperatures but very high stress levels are required and if present would be linked to crack tips, particularly where a thin ligament remains, i.e. close to the free surface. Hence is it considered that the cracking observed in weldments occurred almost exclusively during boiler construction and any subsequent growth, if present, will have been minor.

5.2. Implications for Repair Strategy

One feasible route for returning the defective boilers to service would be to effect weld repairs with post weld heat treatment. Since the defects in the boilers of Reactor 2 are diagnosed to arise during the stress relief heat treatment following welding there is concern that this repair option could lead to a re-occurrence of the problem. However, the results of the current investigation, and an understanding of the cracking mechanisms, provide confidence that repairs could be successfully undertaken if certain precautionary and ameliorative measures are employed. This confidence is reinforced by the fact that the cracked plates are not atypical and that the vast majority of plates were satisfactorily welded and stress relieved during the original fabrication of the boiler shells.

Clearly, the first requirement would be to remove cracks and any damaged material. Major defect removal should be straightforward and can be confirmed by magnetic particle inspection. Complete removal of cavitated material, is likely to be a more intractable problem. Further sampling, described in detail elsewhere (19), has shown that cavitation damage and micro-cracks in coarse grained HAZ extend over a considerable circumferential distance beyond the major crack tips. Indeed, in the boiler 2C 6/7 weld this damage was encountered around almost the entire weld circumference. In the through thickness direction and outwith the major defects, the damage appears to be limited to a depth of < 20mm although this would need to be confirmed by replication. Examination of samples from within the major defects indicates that when the cracks are > 15 mm in depth the cavitation ahead of the tips is limited to a few mm. It is important that cavitation damage is removed or reduced to a low level so that only discrete cavities remain; a criterion should be set. This is essential to ensure remaining damage does not develop during the welding and heat treatment cycle. In addition there will be a need to remove previous weld repairs and remnants of attachment welds lying in close proximity to the intended weld repair preparation.

Since it has been demonstrated that coarse grained HAZ is the more susceptible to stress relief cracking its presence in the repair welds should be largely avoided by control of the weld deposition process to optimise the microstructure and reduce residual stresses, Figure 9(a). A multi-layer grain refinement technique can be employed which ensures that the coarse grained HAZ of each weld bead is refined by the subsequent overlying bead. This process may reduce residual stresses, but may not entirely eliminate coarse grained HAZ, although it would minimise the risk of any significant stress relief cracking. Since a weld repair without heat treatment would leave higher residual stresses a post weld heat treatment would be more appropriate. However, both welding induced residual stresses and thermally induced bending stresses generated during the heat treatment would need to be controlled. In particular, the depth of the weld repair preparation may need to be optimised to ensure that not only is cavitated material removed, but thermally induced bending stresses acting on the original coarse grained HAZ on the floor of the preparation are minimised. Since there is a possibility that hydrogen cracking acted as an initiator to the observed stress relief cracking strict controls should be employed in the repair welding to prevent its occurrence.

It is apparent from the cracking observed that at the original stress relief temperature of 600 °C the coarse grained HAZ of BW87A possessed low creep ductility. In hindsight this temperature is too low to adequately stress relieve and/or temper the HAZ and, indeed, Babcock and Wilcock subsequently recommended a temperature of 650 °C. Consideration needs to be given to establishing a more suitable post weld heat treatment temperature which would confer greater creep ductility in the coarse grained HAZ. Although the repair weld HAZ may contain little coarse grained material this microstructure will still be present in the original weld beneath any repair preparation and in the axial welds which intersect the repair. If all of the factors outlined above can be taken into account it is considered that weld repairs can be successfully accomplished.

6.   ACKNOWLEDGEMENT

We would like to thank our many colleagues at the Berkeley Centre and at site who have contributed to this work, particularly Mr I Lingham (Morson). This paper is published with the permission of the Director Technology and Central Engineering, BNFL Magnox Generation.

7.   REFERENCES

(1)   West of Scotland Iron and Steel Institute, Special Report on Failure of a Boiler during a Hydrostatic Test at Sizewell Nuclear Power Station, 1964.

(2)   F M Burdekin and T Boniszewski, Preliminary Repair and Investigation into Failure of Boiler at Sizewell, BWRA Confidential Report, LD 1481/1/63, 1963.

(3)   'Magnetic Particle Inspection Results for Circumferential Weld between Courses 7 and 6 at Drop 'B', Boiler 2C' Nuclear Electric Report, SIZA/95/BLR2C/31, 1996.

(4)   D A Wood, Sizewell A Boiler 2C Circumferential Weld Joining Courses 6 and 7, Nuclear Electric Report, SIZA/95/BLR2C/83, 1996.

(5)   D A Wood, Sizewell A Boiler 2C Circumferential Weld Joining Courses 5 and 6, Nuclear Electric Report, SIZA/95/BLRC/84, 1996.

(6)   P E J Flewitt and R K Wild, Physical Methods for Materials Characterisation, IoP Publishing (Bristol), 1994.

(7)   R V Maskell, Air Oxidation Kinetics on Ferritic Steels - Compilation of Published Data, Babcock Power Research Report, (08)/82/48, 1983.

(8)   L F Exworthy, I J Lingham and B J C Ellis, Metallurgical Examination of Samples Extracted from Circumferential Boiler Shell Welds, Sizewell A Power Station, Magnox Report, TE/SXA/REP/0129/97, Issue 1, 1997.

(9)   J MacMillan and P E J Flewitt, Micron, 6, 141, 1975.

(10)   D Lonsdale and P E J Flewitt, Materials Science and Engineering, 41, 127, 1979.

(11)   D S Wilkinson, K Abiko, N Thyagerajan and D P Pope, Met. Trans., 11A, 1827, 1980.

(12)   M Lamb, Thermal Embrittlement of Boiler Shells, Sizewell A Power Station, Magnox Report, TE/SXA/REP/0057/97, 1997.

(13)   A Dhooge and A Vinckier, Int. J. Pres. Vessels and Piping, 27, 239, 1987.

(14)   J W Hancock, Metal Science 10, 319, 1976.

(15)   D A Miller and T G Langdon, Scripta Metall., 14, 179, 1980.

(16)   R E Dolby and G G Saunders, A Review of the Problems of Reheat Cracking in Nuclear Vessel Steels, The Welding Institute, Report No. 18/1976/M, Abington, 1976.

(17)   T Naiki and H Okabayashi, J. Japanese Welding Soc., 39, 1059, 1970.

(18)   F Coe, Welding Steels Without Hydrogen Cracking, Welding Inst. Publication, 1973.

(19)   E McDonald et al, Paper 6, these Proceedings.

Table 1: Ultrasonic inspection results showing sizes of major defects in circumferential welds: Boilers 2A, 2C and 2D.

Boiler	Weld/ Toe	Start (mm)*	End (mm)*	Length (mm)	Depth (mm)		Comment
					Best Estimate	Upper Bound	
2A	6/7 top toe	9580	10360	780	33.3	35	minor/shallow indications beyond one end
2C	5/6 top toe	4223	5173	950	13	16	intermittent at ends
	6/7 top toe	2910	7320	4410	24.2	26	defect shallow and intermittent over first 865 mm
	6/7 bottom toe	3493	4783	1290	25.5	27	intermittent defects beyond both ends
2D	6/7 top toe	7450	9405	1955	21	23	intermittent and shallow indications at both ends
	6/7 bottom toe	7465	9500	2035	22	24	intermittent and shallow indications at both ends

* - Datum taken as centre-line of axial weld 7a and measurements increase positively in a clockwise direction with gas flow.

Table 2. On site hardness measurements (Hv) on boiler shell welds

Weld Identity	Measurement Location	Hardness Hv			
		BLR 1A	BLR 1B	BLR 1C	BLR 1D
7/Top Dome	top dome	250	191	215	177
circumferential	HAZ	276	210	235	191
	weld	236	241	250	246
	HAZ	236	201	248	220
	course 7	192	186	270	199
Axial 7A	plate 7C	186	166	133	194
	HAZ	202	185	263	271
	weld	264	242	263	244
	HAZ	204	164	262	223
	plate 7B	185	156	196	197
Axial 7B	plate 7A	176	188	228	199
	HAZ	308	240	253	208
	weld	245	247	287	256
	HAZ	187	233	260	263
	plate 7C	177	245	254	217
Axial 7C	plate 7B	176	188	228	199
	HAZ	308	240	253	208
	weld	245	247	287	256
	HAZ	187	233	260	259
	plate 7A	177	245	254	175
Axial 6A	plate 6C	174	166	196	206
	HAZ	213	239	159	280
	weld	241	225	232	259
	HAZ	274	204	221	259
	plate 6B	203	196	186	175
Axial 6B	plate 6A	178	176	182	233
	HAZ	209	226	270	249
	weld	247	216	236	240
	HAZ	250	186	250	253
	plate 6C	187	161	207	209
Axial 6C	plate 6B	191	221	187	159
	HAZ	193	286	213	265
	weld	236	242	253	257
	HAZ	196	234	265	264
	plate 6A	193	189	189	194
6/7	course 7	202	142	259	202
circumferential	HAZ	256	212	283	258
	weld	250	239	210	245
	HAZ	207	286	256	256
	course 6	204	196	194	203

Table 3. Summary of Examination of scoop samples

Boiler location	Relation to major crack	HAZ grain size (µm)	Max. HAZ hardness (Hv) (10kg)	Plate grain size (µm)	Plate bulk hardness (Hv) (10kg)	% Bainite in plate
2A-6/7-7AY	AMC	49	309	22	212	21
2A-7A-Plate	-	-	-	22	208	21
2A-6/7-7BX	UCP	49	327	28	198	37
2A-6/7-7CX	UCP	140	333	32.4	207	38
2A-6/7-6AX	UCP	49	336	46.5	226	40
2A-6/7-6BX	UCP	49	373	44	246	43
2A-6/7-6CX	UCP	49	304	26.4	207	44
2V-6/7-7CY	AMC	24-37	285	25	227	58
2C-7A-Plate	-	-	-	25.4	189	24
2C-7C-Plate	-	-	-	14.2	196	28
2C-6/7-6CY	AMC	98	292	29.5	212	29
2C-6C-Plate	-	-	-	26.2	220	25
2C-6/7-7AX	UCP	35	304	33	211	53
2C-6/7-7BX	UCP	70-98	314	33.2	222	35
2C-6/7-6AX	-	49	344	28.6	214	26
2C-6/7-6BX	UCP	70	311	28.6	202	23
2C-6/7-6CX	CPR	140	301	29.5	235	29
2C-5/6-5CX	UCP	49	322	38.4	220	35
2C-5/6-5AX	UCP	49	306	24.3	204	40
2C-5/6-5BX	UCP	24.5	285	28	195	30
2D-6/7-7AX	CPR	98 to 140	317	40.5	212	25
2D-6/7-7BX	UCP	98	373	33.4	235	56
2D-6/7-7CX	UCP	49	330	32.7	206	37
2D-7A-Plate	-	-	-	40.5	205	25
2D-6/7-6AY	AMC	140	294	32.2	227	47
2D-6/7-6AX	CPR	98 – 140	311	42.1	215	45
2D-6/7-6BX	UCP	131 – 140	290	22.6	185	22
2D-6/7-6CX	UCP	70 – 98	376	28.8	212	19
2D-6A-Plate	-	-	-	17.9	201	27
2D-6/7-7AX Internal	CPR	98 – 140	325	40.5	219	25
2D-6/7-6AX Internal	CPR	198	340	17.9	236	27
2D-5/6-6AX Internal	UCP	98	328	17.9	230	27
2D-5/6-5AX Internal	UCP	98	290	-	211	-

[Cylindrical butt welds are designated X/Y, where X and Y are the numeric identification of joined courses; a sample designated 2C – 6/7 7A originates from the course 7 side fusion boundary in plate A of circumferential weld 6/7 on boiler 2C. Generally samples are centred on weld HAZ, some are taken from plate, '-Plate'. The 'Boiler Location' description is followed by X or Y; Y indicates scoop taken ≤ 200 mm of a crack tip in the same plate, and X was either in cracked plate ≥ 2m from crack tip or in uncracked plate.]

UCP = Uncracked plate : CPR = Cracked plate remote from main crack :

AMC = Adjacent to main crack

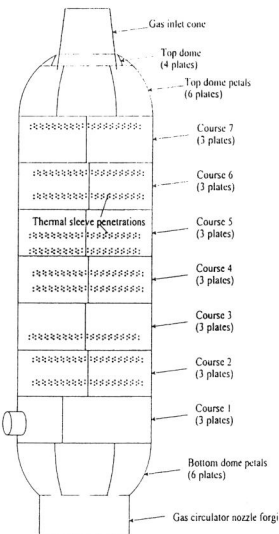

Figure 1: Schematic diagram
showing boiler construction.

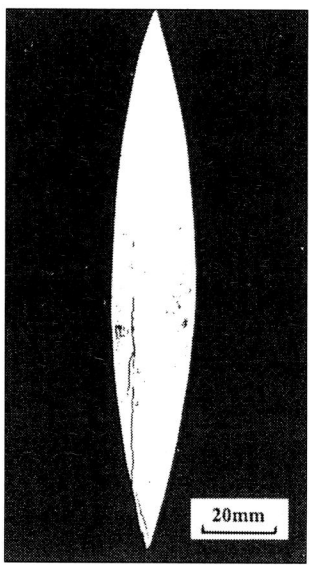

Figure 2: Cracking in extracted boat sample
on course 6 side of 2C-6/7 as revealed by
MPI; viewed from top surface.

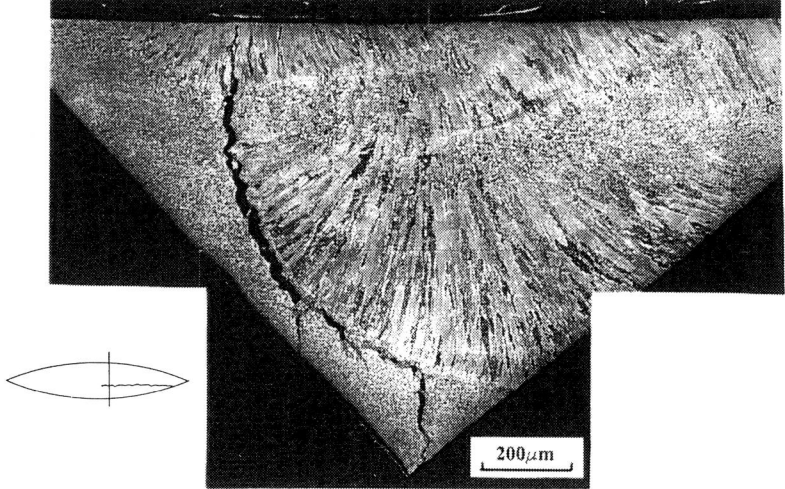

Figure 3: Cracking in transverse section through boat sample in Figure 2.

Figure 4a: Coarse grained HAZ of weldment showing intergranular cracking.

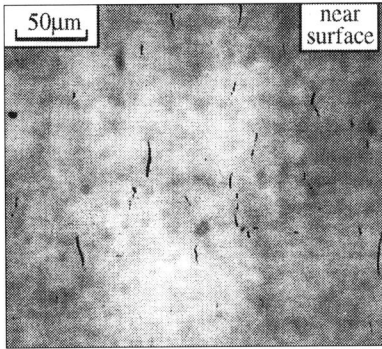

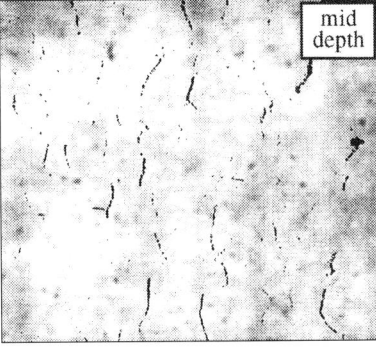

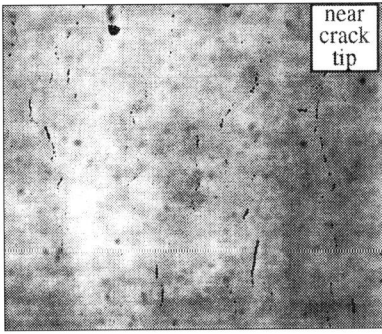

Figure 4b: Cavitation and subsidiary cracking in HAZ of weldment.

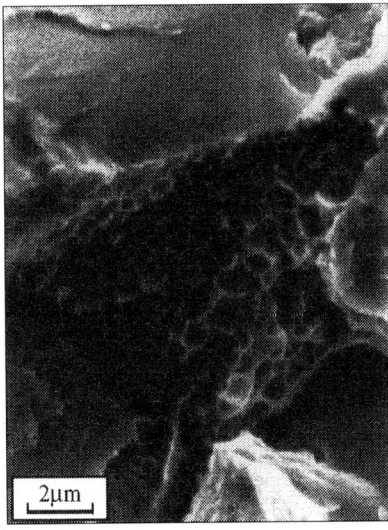

Figure 5: Secondary electron image of the fracture surface showing a cavitated grain boundary.

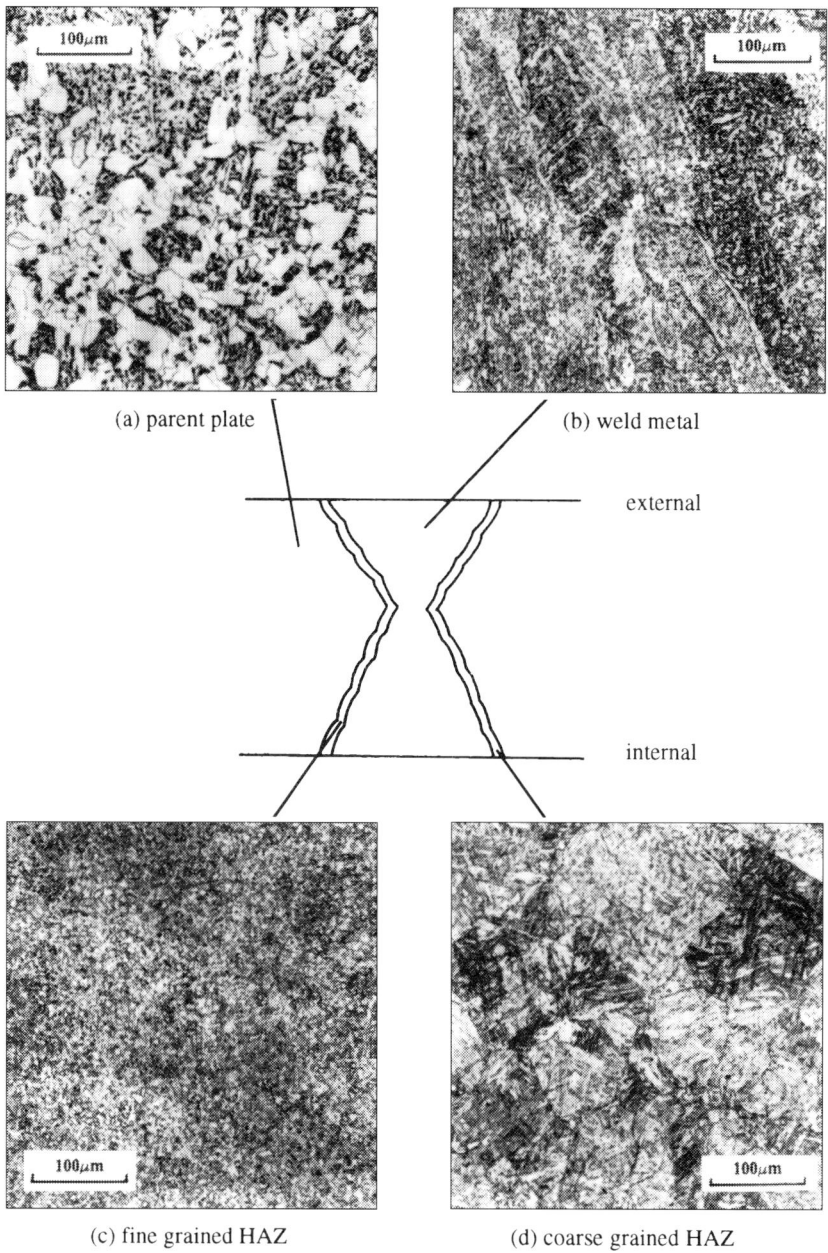

Figure 6: Optical micrographs of the key microstructures for a boiler weldment (a) plate (b) weld metal (c) fine grained HAZ (d) coarse grained HAZ.

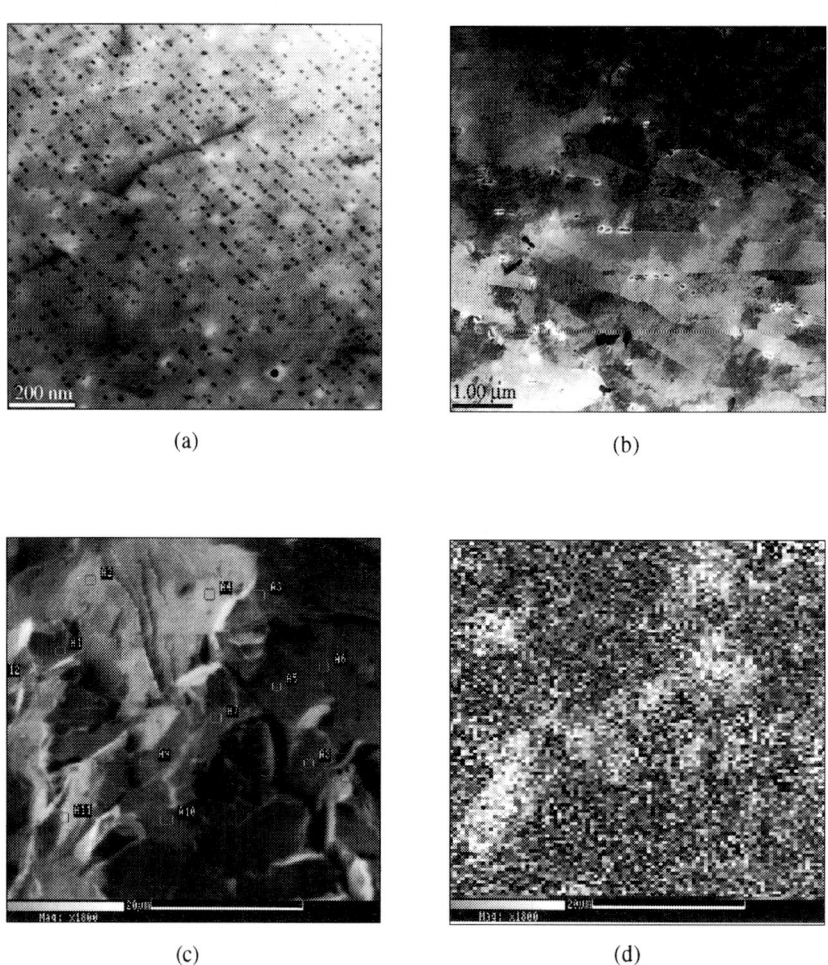

(a)

(b)

(c)

(d)

Figure 7: Transmission electron micrographs showing (a) aligned MC type carbide precipitation and (b) bainitic lath structure typical of HAZ. A secondary electron image of an Auger fracture surface is shown in (c) and a AES phosphorus map of the same area in (d).

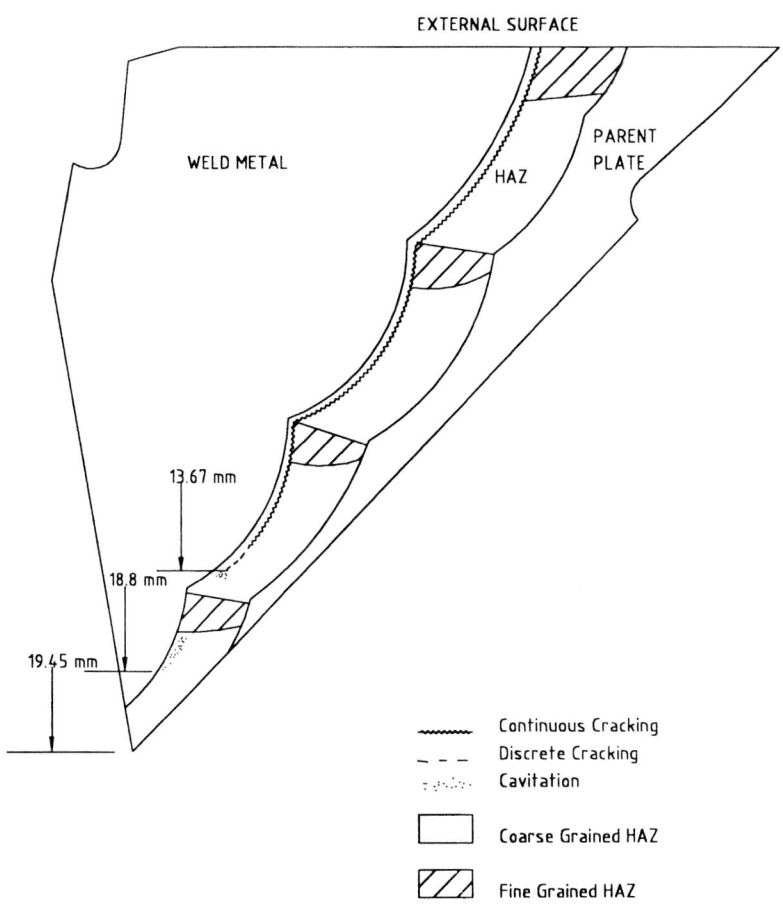

EXTERNAL SURFACE

WELD METAL

PARENT PLATE

HAZ

13.67 mm

18.8 mm

19.45 mm

Continuous Cracking
Discrete Cracking
Cavitation
Coarse Grained HAZ
Fine Grained HAZ

Figure 8: Schematic diagram showing distribution of coarse and fine grained HAZ, cracking and cavitation in boat sample 2C-6/7-7C.

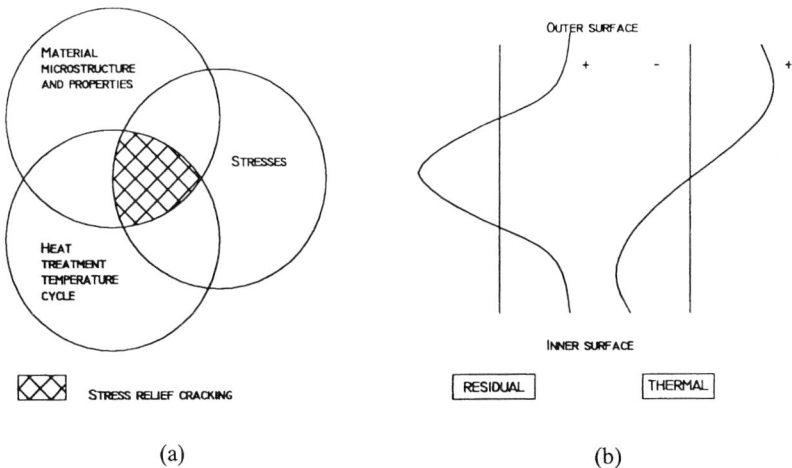

Figure 9: Factors contributing to stress relief cracking and its distribution in weldments (a) Venn diagram of factors leading to cracking and (b) profiles arising from residual and thermal stresses.

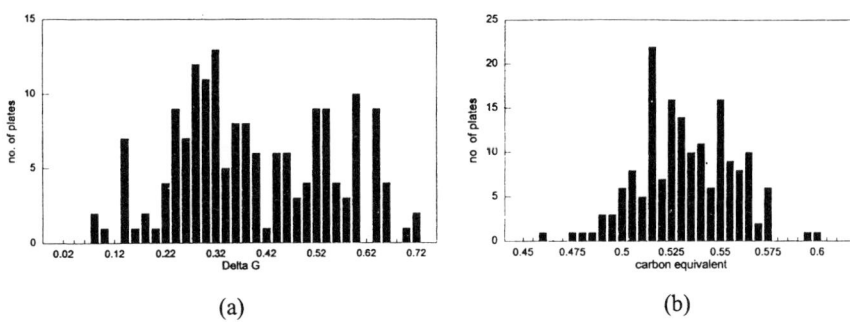

Figure 10: Susceptibility of R1 and R2 boiler plates to (a) stress relief cracking given by ΔG and (b) hydrogen cracking via carbon equivalent, CE.

S690/003/99

Repair options for defects found in boiler shells

G S ANDERSON
BNFL Magnox Generation, Berkeley, UK
W J G LITTLE
Mitsui Babcock Energy Limited, Glasgow, UK

ABSTRACT

During inspection of the circumferential weld seams on boiler shells of Sizewell 'A' Reactor 2, defects were found which were greater than those considered in the boiler shell safety case. It was necessary to either justify the presence of defects or carry out repairs to reinstate the circumferential weld seams. Leaving the defects in the boiler shell would have require protracted justification.

Various repair options were considered and these are described. However following a detailed assessment of the post weld heat treatment of a weld repair it was concluded that the repair option was technically feasible, economically viable, and would result in consent for continued operation of the plant.

1. INTRODUCTION

Following the discovery of defects in the shells of boilers 2A, 2C and 2D of Sizewell A Power Station, a review was carried out as to the best approach to ensure that the station was returned to service in an appropriate timescale and with a high margin of safety.

The review was not constrained in any way and a wide range of options were considered. These ranged from the concept of leaving the defects in the shells without repair to replacement of the entire boiler shells. The options were developed and reviewed by a multi-disciplined team of engineers and metallurgists to ensure that each aspect of the proposed options could be fully evaluated. The team consisted of staff from BNFL Magnox Generation (Magnox) who were familiar with the plant and the current safety issues and staff from Mitsui Babcock Energy Limited who were experienced in design, welding and plant refurbishment. The range of skills and different backgrounds of the staff involved from both companies were blended into a team which was ideally suited to undertake the review of the repair options and select the most appropriate option for further development to a full repair scheme

This paper describes the options considered, presents the work carried out on each option, and provides justification for the option selected.

2.   EVALUATION OF THE DEFECTS

The original Boiler Shell Safety Case for the boilers on Sizewell A Power Station considered the integrity of the boiler shell under various combinations of loading:

- Pressure

- Self weight

- Thermal transient during start-up

- Thermal transient during shut down

- Seismic

Stresses in the shell were determined for each of the above mechanisms at the various main seam locations in the boiler shell. From this information and from the materials properties contained in the materials property database held by Magnox, an evaluation of the margins of safety were determined for the reference defect size. A surface breaking defect of 15 mm (the reference defect) had a reserve margin in excess of 2.0. The defects found were well outwith this established position and therefore the original boiler shell safety case was invalid. Possible ways forward were identified at this stage. One was to review the safety case for continued operation of the boilers with the defects present in the shells and the other was the development of repair options.

3.   CONTINUED OPERATION OF THE PLANT WITH THE DEFECTS

The existing safety case was reviewed in terms of the critical defect size for the surface breaking defects which were present in the boiler shell. The analysis was carried out using actual defect lengths and depths as determined by the NDT inspections. In addition to the operating and self weight stresses, seismically induced stresses and estimated values for residual stresses were considered in the fracture assessment. Fracture toughness values were assumed to be in line with those specified in the Magnox materials database with a minimum value of 110 MPa$\sqrt{m}$ being used.

For the defects in seam 6/7 of boilers 2A, 2C and 2D, the lowest resultant reserve factor was 1.5 based on a fault pressure of 2.0 MPa. Sensitivity studies showed that for normal operation and for the temperature fault case, reserve factors were higher. As part of the sensitivity analysis carried out, the worst possible defect size was calculated taking into account possible NDT errors and this reduced the reserve factor from 1.5 to 1.4. In order to give further assurance, surface residual stresses were measured and this confirmed that the surface residual stresses were less than the values used in the assessments.

The assessment carried out incorporated all the primary and secondary loadings which the boiler shell could experience and demonstrated that for the defects found, the lowest reserve factor was 1.4. While a safety case could have been made for a return to power with the defects in place Magnox Electric intended to operate Sizewell A beyond its design life, providing justification for this via the periodic Safety Review. With such a lifetime extension in mind, it was decided not to make a case for leaving the defects in place and as a consequence consideration was given to the repair option.

As part of an interim strategy to return the plant to service while the feasibility of a repair was being evaluated, consideration was given to operating the plant at reduced pressures. This would have involved re-setting the safety valves to achieve fault pressure of 1.85 MPa with a corresponding reduction in operational pressure. This increased the defect size reserve factor to approximately 1.75 for the measured defect sizes with a slightly lower value for the worst case NDT defect sizes. However, this approach would have require extensive justification effort and protracted clearance with the regulator. It was decided that in view of the proposed timescale for the repair, this effort was better spent progressing the repair work and ensuring a timely and successful implementation of the repair.

4.   REPAIR OPTIONS

The criteria established for the repair of the boiler shells were:

- The repair had to be of such integrity that a robust safety case could be made to ensure that a license instrument for continued operation of the plant would be granted.

- The repair had to be economically justifiable against the revenue over the future operational life of the plant.

- The repair had to be conducted in such a timescale that a sufficient period of operation would remain prior to the current 'end of station life'.

The various options for the repair of the boiler shell considered were:

- Welded repair with post weld heat treatment.

- Welded repair without post weld heat treatment.

- Machining to remove the defects.

- Mechanical reinforcing of the boiler shell.

- Total replacement of the boiler shells.

4.1 Weld Repair with Post Weld Heat Treatment – Initial Assessment

It was recognised that heat treatment had been the cause of the original cracking and while the reasons for this were known (1), there was considerable concern as to the effect of a further heat treatment on the boiler shell and ancillary components. It was recognised that a fully circumferential band of shell would have to be heated to approximately 650°C and that the heat distributions would have to be controlled over several meters along the shell. Problems of installing internal insulation to protect the tubebank, deloading the boiler shell and the possibility of cracking other welds during the heat treatment meant that risks were perceived to be high. However no detailed engineering assessment had been carried out to either quantify or reduce the perceived risks before this option was abandoned.

4.2 Mechanical Reinforcement of the Boiler Shells

As the axial stresses in the boiler shell were low, mechanical reinforcement of the boiler shells by some means was considered. The design principles for such a scheme were:

1. To provide an additional axial load path which would result in a significant reduction in axial stresses at the defect site.

2. The fixing system was to be based on bolting rather than welding.

3. Ensure the adequate integrity of the boiler shell with the reduced axial load on the defects and the presence of holes in the shell.

Plates capable of carrying axial load would be bolted to the shell above and below the defects. The temperature of the plates could be raised prior to installation so that when the plate and shell were at the same temperature, the plate would be under tension and the shell under low compressive strain. On the pressurisation of the boilers, the reinforcing plate and the shell both would carry tensile load.

Because of the length of the defects, it would not be possible to accurately form a plate to the curvature of the shell and any gaps would cause very high loads in bolts used to connect the plates to the shell. As a result, a series of straps was designed which were sufficiently narrow to allow machining of the shell into a series of flats against which the straps could be bolted. This had the added advantage of providing a flat area on both the straps and the shell for frictional transfer of loads between the shell and the straps. This removed the need to use fitted bolts through the shell as high strength bolts could be used to provide the necessary clamping force. Ideally, straps would be fitted to both sides of the boiler shell, but the presence of internal circumferential baffles precluded the use of full length straps on the inner surface. As a result, only an external strap was possible and this not only limited the amount of load which could be carried, but also introduced a small element of bending into the shell due to the change in direction of the load path. A schematic of the proposed arrangement is shown in Figure 1. As can be seen the high strength bolts went through the shell and were sealed using metal O rings to prevent gas escape.

The major drawbacks with the scheme were:

1. The additional bending moment which resulted from the change of the plane of the axial load.

2. The necessity to monitor the load being carried in the straps at all times to ensure satisfactory transfer of load from the defective area.

3. The need to remove and replace the mechanical straps during weld re-inspections.

4. The integrity of the pressure seals during operation of the plant.

It was felt that while the integrity of the connection to the shell could have been proven, the problems of installation, sealing and monitoring to ensure reliable load carrying were such that acceptance of this scheme would be difficult. There was also the issue of drilling a large number of holes through the shell.

It was concluded that it was unlikely that the numerous technical issues could be resolved to the extent necessary to produce a robust safety case and, as a result, the mechanical reinforcement of the boiler shell was abandoned.

4.3 Replacement of the Boiler Shells

The boiler shells were fabricated in Dalmuir Works, Scotland and were delivered to site as individual courses (cylinders) typically 8 ft. long which were then joined by circumferential welding to form complete boiler shells. The completed welds were locally heat treated. The boiler shells were then set in their final position on the site and the associate ductwork and pipework connected. The tubes were subsequently inserted element by element. It was possible therefore to reverse the procedure, dismantle the boilers and remove them for site. The low radiological dose levels in the boilers at Sizewell A meant that if new boiler shells were made, they could be refitted with the existing boiler tubes, drums and connecting pipework.

While this approach was technically feasible and the manufacturing facilities existed to fabricate new boiler shells, the timescales and costs for such an operation were prohibitive – a budget estimate of £25 million per boiler was produced. The option was, therefore, considered not to be economically viable.

4.4 Machining of Boiler Shell to Remove the Defects

As all the defects were surface breaking, it would have been possible to machine the surface of the shell to remove them. The main constraint was the minimum remaining ligament required to sustain the pressure envelope - the assessment criteria for the remaining ligament would set the maximum depth of defect which could be excavated. The boiler shells were originally designed to BS.1500. Consideration was given to carrying out the design of any modification to the modern equivalent BS.5500. However, as this is a non nuclear code, any design of plant modifications was carried out to ASME. The criteria established were:

1. The resultant stresses due to pressure and thermal load cases must meet the requirements of ASME III.

2. The maximum peak stress at any discontinuity should not exceed yield level for the material and temperature.

3. The excavation should not result in the remaining shell thickness being less than that required by the appropriate code (in this case ASME III).

4. Any localised stress concentrations should be well removed from the weld HAZ.

5. The shape of any excavated area should not impair the ability to inspect the weld.

6. The excavation method should not introduce significant surface residual stresses.

These were conservative criteria and ones which would stand the test of scrutiny in any safety case. The minimum ligament thickness criteria meant that the excavation depth could not exceed 16 mm and thus only the seam 5/6 defect on boiler 2C could be removed in this manner. Since the 6/7 defects on all boilers exceeded this depth, this option was clearly not viable for all defect locations, and so other options needed to be considered. However, extensive 3D finite element analysis was still required to demonstrate that the criteria had been fully satisfied. Details of the excavated profile at seam 5/6 is shown in Figure 2

The actual machining operation for the 5/6 defect required significant development. A large mock up was fabricated in carbon steel plate to simulate a 3 meter high section of the boiler shell. Much of the machining development work was carried out on this mock up to set the cutting speeds and the method of profiling the final shape. Once the machining methods had been established on the curved carbon steel mock up, a section of BW87A plate was inserted into the mock up to assess the machining characteristics of the boiler shell plate.

Experimental measurements of residual stress after machining had shown that the low cutting speeds used generated low residual stress. It was found that these could be further reduced by hand finishing the profile using a 60 grit disc. By removing 1 mm of material in this manner, surface residual stresses were reduced to less than 20 MPa which provided margin relative to the criterion 6 above.

4.5 Weld Repair of the Defective Seams without Post Weld Heat Treatment
The criteria for the weld repair of the boilers without heat treatment were:

1. A fine grained HAZ must be achieved with good fracture toughness.

2. The level of residual stress produced must be such that the critical defect size was well within the capability of NDT techniques.

3. The repair should be carried out to an appropriate code and be code compliant.

4. The structure must be code compliant after the completion of the repair weld.

5. The weld procedure must be qualified to an accepted code.

The weld repair development work carried out for the non post weld heat treated repair was similar to that reported for the post weld heat treated repair (2) and will not be repeated here. It is sufficient to state that the weld characteristics developed were such that they would have proven acceptable for a non post weld heat treated repair. The key feature which determined the acceptability of this approach was the demonstration that the residual stress distributions could be predicted for the repair of the boiler shell and that the levels were such that the critical defect sizes were within the capability of NDT inspection techniques.

The most relevant code for the repair of nuclear power plant components, either with or without post weld heat treatment, is ASME XI. Under the criteria set down in ASME XI, a non post weld heat treated repair is permissible providing:

1. The repaired depth does not exceed 50% of the plate thickness.

2. A temper bead repair technique is developed and validated.

3. The repair is reassessed to either the original code of construction or current equivalent code.

The thickness of the Sizewell A boiler shells is 57 mm, and the code minimum thickness for the shell is 41 mm. The maximum depth of the defects Boilers 2C and 2D was 27 mm, leaving at least 30 mm of sound material. For Boiler 2A, the defect was up to 35 mm deep. A possible way forward was to machine down the area of the weld to a 42 mm thickness local to the repaired area. This would have left defects on Boilers 2C and 2D with repair depths below the 50% ASME XI limit, whereas Boiler 2A would have been outwith the code requirement. For this reason and to minimise the volume of weld metal to be deposited, the options for a repair with a locally reduced wall thickness and a full wall thickness were assessed. The two variants are shown in Figure 4.

4.5.1 Review of Residual Stress Levels Associated with Weld Repairs

The residual stress levels and their distribution through the thickness of a repaired component are used, in conjunction with the material properties and the operational stresses acting on the component to determine critical defect sizes using the R6 procedure. Associated with the R6 procedure is a Compendium of Residual Stress Profiles (3) which defines the residual stress distributions for various types of welds including a repair weld in a double V preparation plate butt weld. However, the geometry detailed for this repair is different from that projected for the Sizewell A boiler shells, - the former is a short excavation local to one side of a weld HAZ whereas the latter is a large excavation of the entire original weld width over a length of several meters. Despite the differences, the distribution recommended in the Compendium, and shown in Figure 3, was used to give an indication of the critical defect sizes. An R6 assessment was carried out using the distribution shown in Figure 4 and using the upper bound material/weld yield stress of 500 MPa. For a full thickness shell this gave a critical surface breaking defect size of just under 4 mm. For the shell which had been machined to minimum thickness the critical defect size was closer to 3mm. Clearly these defect sizes were too close to the minimum defect size which could reliably be detected by NDT (3mm) and a sufficiently robust safety case could not be formulated.

A review of other literature demonstrated that the residual stress distribution in the Compendium could be very conservative and cases by Mathieson (4) and Leggat (5) indicated that a realistic residual stress distribution would be more akin to that shown in Figure 5. With such residual stress distributions, the critical defect sizes would be significantly greater. A programme of work was therefore initiated to demonstrate the residual stress magnitudes and through wall distributions in repair welds of the type to be used on the Sizewell A boiler shells.

4.5.2 Computation of Residual Stresses in Proposed Repair Weld

In the development of the programme to measure or compute the residual stresses in repair welds, it was recognised that it would not be possible to determine the residual stresses by purely experimental means. This is because the stiffness of the boiler shell could not be modelled in small test plates and a full size mock up several meters long and approximately half of the boiler shell diameter would be required to model the constraint effects. As a result, a combined experimental and analytical programme was developed. Early development work was carried out using Ducol BS.1501-271B material until the specially cast BW87A replication material became available. The main steps in the programme are listed below:

- Fabricate test plates in Ducol material and measure surface and through thickness residual stresses.

- Carry out stress strain material tests for plate and weld metal at 20-800°C.

- Use Finite Element Analysis to predict residual stresses in the test plates.

- Compare the measured and predicted values.

- Fabricate test plates in BW87A material and measure surface residual stresses.

- Use Finite Element Analysis to predict residual stress levels in the BW87A test plates.

- Compare the experimental and predicted values for the BW87A test plates.

- Predict the residual stress distributions in the boiler.

The test plates used for the above work were those used for the development of the repair weld procedure and for the weld procedure qualification work. The test plates were fabricated using the original boiler shell butt welds process, with double sided V prep butt welds with heat treatment to 600°C. Following this process, backing plates were welded to the test plates to provide a degree of restraint in excess of that imposed by the boiler shells. Grooves were then machined along ¾ of the length of the test plates to simulate the repair weld preparations. This allowed the geometry of the end of the repair weld to be accurately represented in the test plate.

The repair weld was laid down using a four layer technique, as described in (2). Measurements of residual stress were taken at various positions across the weld and at the end of the weld using the centre hole drilling technique. The results for the Ducol test plate showed that the outer surface transverse residual stresses are of yield magnitude. However, the levels at the end of the weld are significantly lower at the surface, probably due to the fact that the depth of the repair weld decreases towards the end of the repair excavation. Surface residual stresses were also measured on the BW87A test plates using the centrehole strain gauge technique. The yield strength of the BW87A material is substantially lower than that of Ducol and this was the reason for the lower measured residual stresses.

It is recognised that the centrehole drilling technique can be subject to error of the order of 16% in areas of yield level residual stress. A second set of residual stress measurements was therefore taken using the X-ray diffraction techniques. These showed good agreement with the centrehole measurements and gave confidence in the results.

In addition to the surface measurements, one set of through-thickness residual stress measurements was taken by Bristol University using the deep hole drilling technique. This again showed good agreement near the surface with the surface residual stress measurements.

The prediction of residual stress levels using Finite Element Analysis techniques is a relatively new application of finite element analysis and, while it has been applied in certain applications in 1997, it did not have the status of a tried and tested analytical method at the time this work commenced. As a result, significant development and validation work was required, to be focused on the specific repair geometry in question. There was limited expertise in this type of analysis and, after a review of capability in both Europe and the USA, it was decided to work with the Battelle Institute from Ohio, USA to predict the residual stress levels in the repair weld.

Discussions took place on the bead deposition sequences and the way the repair was to be made. The first four layers were designed to ensure that the required metallurgical criteria, particularly grain refinement and toughness, were achieved and it was agreed that these metallurgical criteria took precedence over residual stress issues. The remainder of the weld, the 'bulk fill' was then completed in the sequence which gave the minimum residual stresses, providing that this was consistent with the requirements of the welders for producing defect free welds.

The modelling techniques used to predict residual stresses are based on thermo/mechanical analysis with the varying material properties for the range of temperatures applicable to the process being introduced through specialist UMAT subroutines written by the users within the ABAQUS Finite Element package. Details of the analysis techniques have previously been presented by Battelle (6) and will not be discussed in this paper.

Of more immediate interest is the weld bead modelling and results. The actual weld bead pattern, as shown in Figure 6, was idealised to simplify the analysis and reduce the computational time. The resulting finite element idealisation is shown in Figure 7. The first step was to develop confidence that the finite element modelling could indeed give a reasonable prediction of the residual stress levels in the repair weld. Detailed comparisons of the predicted and measured residual stress distributions across the repaired weld are shown in Figure 8 and a comparison of the through thickness predicted and measured distribution are shown in Figure 9. It can be seen that for both these cases very good agreement was obtained.

Further work was undertaken to assess the effect of both parent and weld metal yield strength on peak surface residual stresses. The results showed that the residual stress levels are higher for material with higher yield strength and confirmed that the selection of the A0 type electrode was the most appropriate for this purpose.

4.5.3 Review of Residual Stress Predictive Capability and Effect on Safety Case

The initial work described above demonstrated that the distribution proposed in the R6 Compendium of Residual Stresses (2) was upper bound. A comparison of the residual stress levels as specified in the Compendium compared with those likely to apply to the geometry and materials in question is shown in Figure 10. It can be seen that in the important region near the outer surface, the calculated levels are substantially lower. Using this type of distribution in an R6 assessment would result in a critical defect size of the order of 10 mm.

While this was a substantial improvement over the previous values of 4 mm, it was recognised that there may still be significant difficulties in formulating a robust safety case.

The main difficulties would come from the fact that a repair without post weld heat treatment would be a non code based repair and would rely very heavily on the prediction of residual stress levels by Finite Element Analysis. While it had been shown that there was a capability to predict the residual stress with a good degree of accuracy, the margins using this approach were small, and further significant validation work would be required into issues such as:

- Residual stress distribution at the end of the weld.

- The effect of restraint of the boiler shell.

- Sensitivity of residual stresses to variations in material properties.

- Fracture toughness of existing material at the ends of the welds and at the inside surface of the shells.

It was considered that without further validation work all of the above would reduce the margins on the critical defect size and increase uncertainty to an extent whereby a robust safety case could be expensive and time consuming to formulate. As this was outside the 'economic and timely' criteria set for the repair, work on this approach was discontinued.

5.   POST WELD HEAT TREATED REPAIR

During the course of the various assessments described above, it became obvious that each of the options had difficulties, and none was code compliant. As a result the only code compliant option, the post weld heat treated repair, was reviewed again to ascertain the risks and the likelihood of success.

The key issues were:

- The thermal stresses induced in the shell.

- The support of the tube bank during the heat treatment - the tube bank support brackets would experience temperatures of 600°C, and it was advisable to de-load them.

- The protection of the tube bank from excessive temperatures by installing insulation between the tube bank and the shell.

- The ability of the shell material to withstand a further heat treatment.

- The disconnection from the boiler shell of the gas ducts, the steam drum, the ancillary pipework, etc.

- Loads transmitted through the ductwork during PWHT to the RPV nozzles.

A team comprising both MBEL and Magnox staff was set up and given a ten week period in which to review the issues involved, to assess the level of risk, to produce a project plan with detailed repair schemes for each of the three boilers and to produce a target price for carrying out the repair.

5.1 Shell Stresses

The issue of greatest concern was the axial temperature distributions which would occur in the boiler shell. Analysis showed that if a sufficiently wide band (400 mm) was heated to the proposed heat treatment temperature (650°C) and the temperature was controlled to 250°C over a length of 2 meters either side of the weld, the stresses at the weld would not exceed yield level at the peak heat treatment temperature. The one area of concern was the HP and LP superheater inlet thermal sleeves. These components would be subjected to temperatures of approximately 600°C (the temperature of the original heat treatment when cracking of the circumferential welds had occurred) and were in an area of high stress gradient. Although it had been shown that the thermal sleeves were relatively defect tolerant, a detailed programme of inspections of the thermal sleeves was instigated to confirm their present condition and work was started to develop a repair method in the event that a defect was found which necessitated a repair of a thermal sleeve.

5.2 Tube Bank Support

A review of the tube bank supports confirmed that they would not sustain the weight of the tube bank at the expected temperature of 600°C. It was concluded that it would be possible to support the tube bank from existing lugs welded to the boiler shell dome or from the upper cone assembly using a specially designed support frame. The frame would be man handled through the man access door above the superheater tube bank and bolted together inside the boiler. The support rods and fork ends would be similar to those used in conventional boiler support system which routinely experience temperatures of 360°C in the dead space of conventional boilers. The use of standard and proven components substantially reduced the risk associated with this aspect of the work.

5.3 Insulation

Protection of the tube bank from high temperatures was essential, not only to safeguard the tube bank, but also to limit heat loss. Excessive losses would considerably increase the time to reach the heat treatment temperature and increase power consumption. The use of specially designed panels of Microtherm insulation attached to thin stainless steel sheets would allow the insulation to be pushed between the tube banks and the shell. While many design details had to be resolved, it was agreed that the concept was viable and of a sufficiently low risk that is would not threaten the project.

5.4 Shell Material

Some parts of Boiler 2C had already experienced two heat treatments and trials on material samples demonstrated that further heat treatment at the projected temperature of 650°C would not cause any significant further degradation of the material.

5.5 Disconnection of Associated Components

A review of the practicality of disconnecting items such as the gas ducts, steam drums, downcomers and headers from the boilers was carried out. This confirmed that the work was 'routine engineering' and, with sufficient attention to detail, should not present any serious risk to the project. The RPV nozzles would be protected during the PWHT by separating the duct at the flange. This had been done previously, and so did not pose a serious risk.

All aspects of this repair option were discussed with the NII during the 10 week assessment and, subject to certain principles being adhered to, they foresaw no difficulty in granting consent to operate following a successful repair.

As a result of the reviews, a project plan and costing was developed which demonstrated that post weld heat treatment of the boiler shell was a viable option from economic, licensing and feasibility aspects. It was concluded that the risks involved were well understood and, with sufficient attention to detail were controllable. As a result, the team were able to recommend to the Magnox Board that the repair should proceed.

6.   CONCLUSIONS

The various options for the continued operation of the boilers on Sizewell A Power Station were initially assessed against the safety and integrity criteria for the boiler shell. Where those criteria were satisfied, then further assessments were carried out as to the cost and timescale of the option. The initial proposals for continued operation failed to comply completely with the integrity criteria and there were concerns which could not be satisfied associated with allowing such large surface breaking defects to remain in the primary circuit of a nuclear plant, even at reduced operating pressure.

In assessing the repair strategies, the initial concerns about re-cracking and the complexities associated with post weld heat treated repairs initially seemed insurmountable. However, by the time other options had been evaluated, a better understanding of many of the issues had emerged, allowing this option to proceed.

The work on the machining and non post weld heat treatment options undoubtedly contributed considerably to the re-assessment of the post weld heat treatment case. For example, much of the machining of the weld preparations and associated work developed under the non post weld heat treated option, was used directly for the post weld heat treated case. The machining option was also directly applied to the solution for seam 5/6 for Boiler 2C.

The residual stress predictions for the non post weld heat treated option were invaluable in highlighting the high tensile residual stress which occurred on the inside surface HAZ. This ensured that sufficient care was taken to avoid cracking of those parts of the original weld which were to remain. The residual stress predictions were used, in conjunction with the appropriate creep laws to determine how the strains would accumulate during the heat treatment. The analysis gave confidence that the remaining original weld metal would not crack. Several "what if" cases were run to simulate the heat treatment temperature time profiles to ascertain the best way forward should the heat treatment have to be paused during the cycle.

From the above it can be seen that by addressing the various options in a systematic and thorough manner, much valuable information was gained which helped to formulate the adopted solution. A lesson to be learned from this exercise is the need to thoroughly evaluate all options with detailed engineering studies by a multidiscipline team to evaluate technical risks and develop engineering solutions. By not carrying this out and relying on judgements without having expert input, a viable option was initially ruled out because it 'seemed' too difficult and 'seemed' to carry too much risk.

It is believed that the approach of going through the options in this way has enabled the optimum solution being chosen from a safety, economic and time viewpoint and one which has led to the return to service of the plant.

7.   ACKNOWLEDGEMENT

This paper is published with the permission of the Director Technology and Central Engineering, BNFL Magnox Generation.

8.   REFERENCES

1. Exworthy, L F, Flewitt, P E J & Ellis, B J C "An Evaluation of the Nature and Origin of the Cracking and the Implications for the Repair Strategy", these proceedings.
2. Tolaini, J, "Sizewell A Power Station Reactor 2 Boiler Repairs", Welding Development Report, MBESL Contract No. 984/096/25238, Issue 3, August, 1998.
3. Sanderson, D J, "Recommendations for Revised Compendium of Residual Stress Profiles for R6", AEA Technology Report No. AEA-TSD-0554, June, 1996.
4. Mathieson, P A R, "A Compendium of As-Welded Residual Stress Profiles", Nuclear Electric Report No. TD/SIB/MEM/0233, September, 1991.
5. Leggatt, R H & Friedman, L M, "Residual Weldment Stresses in Controlled Deposition Repairs to 1¼Cr½Mo and 2¼Cr1Mo Steels in Residual Stresses in Design Fabrication and Repair", Proc. ASME Conference, Quebec (PVP-327) (1996).
6. Atluri, S N, Yagawa, K G, "Advances in Computational Engineering Science", ICES 1997 Conference Proceedings, pp.51-56 (1997).

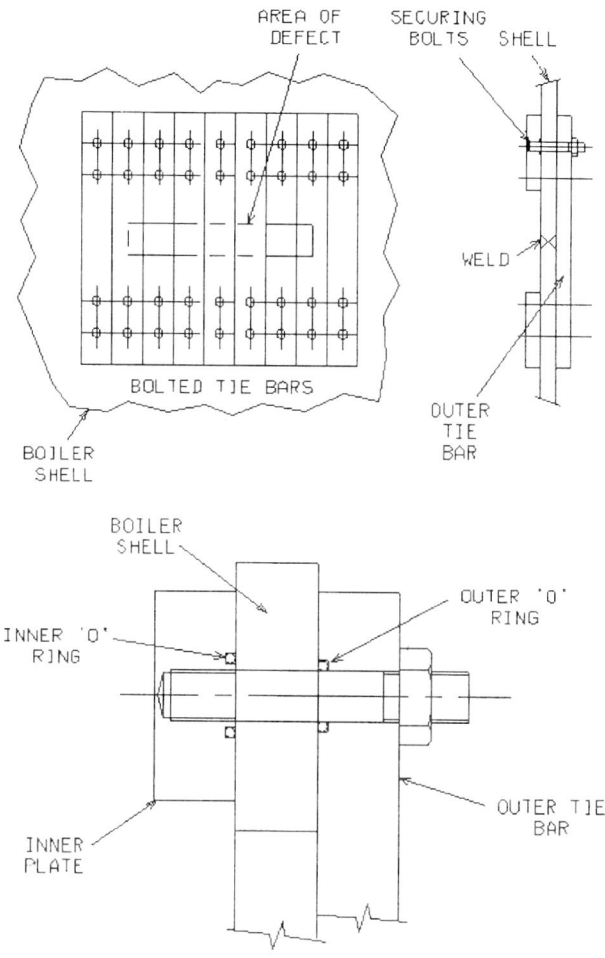

Figure 1: Proposal for mechanically reinforcing boiler shell

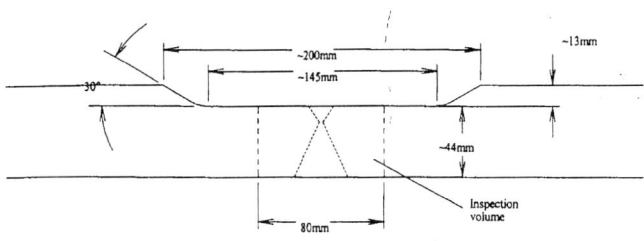

Boiler 2C course 5/6 weld excavation

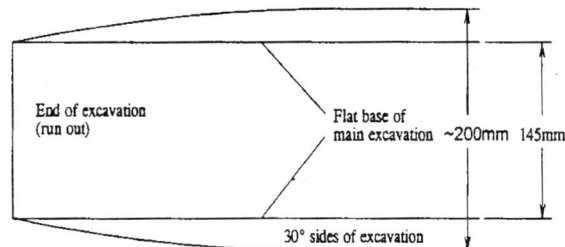

Flat end (run out) – plan

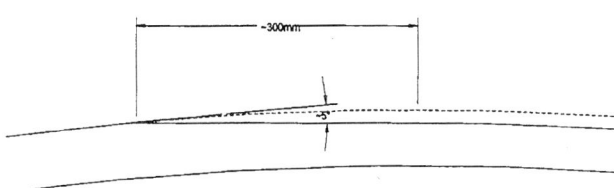

Flat end (run out) – side
elevation

Figure 2: View of excavated 5/6 seam

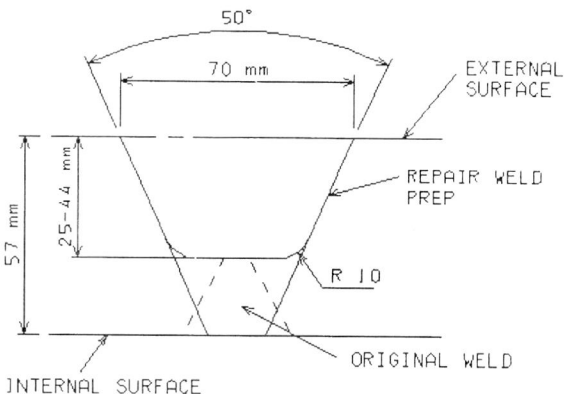

WELD REPAIR EXCAVATION

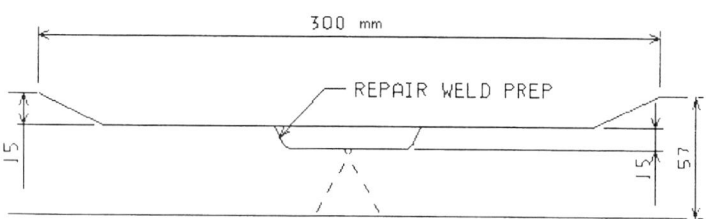

PARTIAL MACHINING OF BOILER SHELL
WITH PART DEPTH WELD REPAIR

Figure 3: Weld preparation excavations

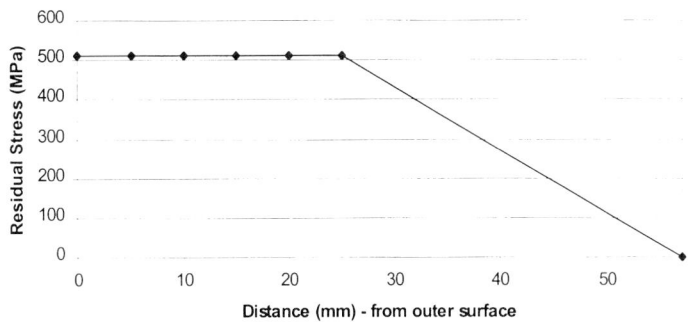

Figure 4: Recommended distribution for residual stress from R6 compendium

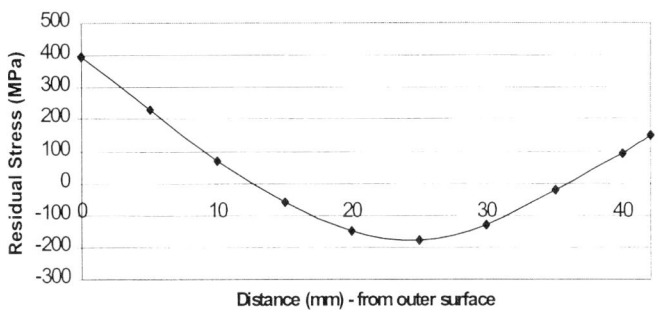

Figure 5: Residual stress distributions in welds Mathieson (Ref#8)

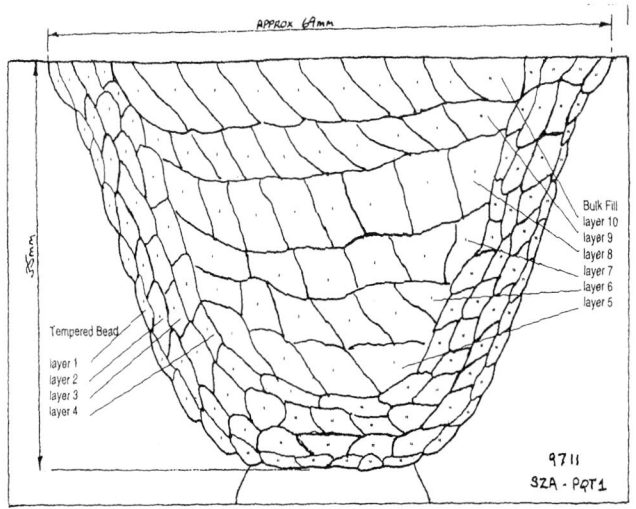

Figure 6: Proposed weld bead deposition sequence

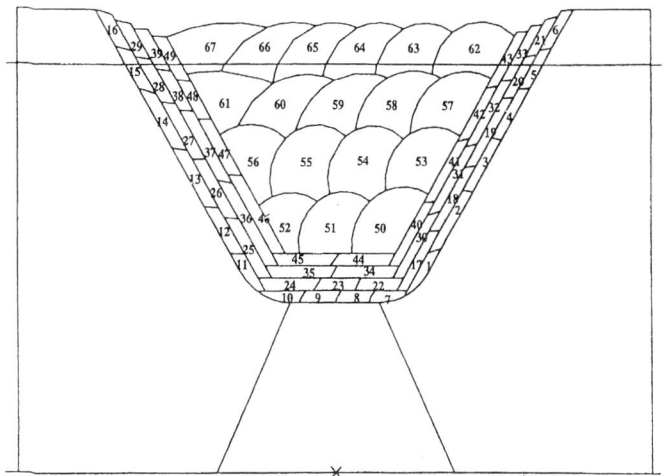

Figure 7: Weld bead deposition sequence used in finite element model

Figure 8: Comparison of Measured and Predicted Residual Stress Distributions Across the Surface of the Repair Weld

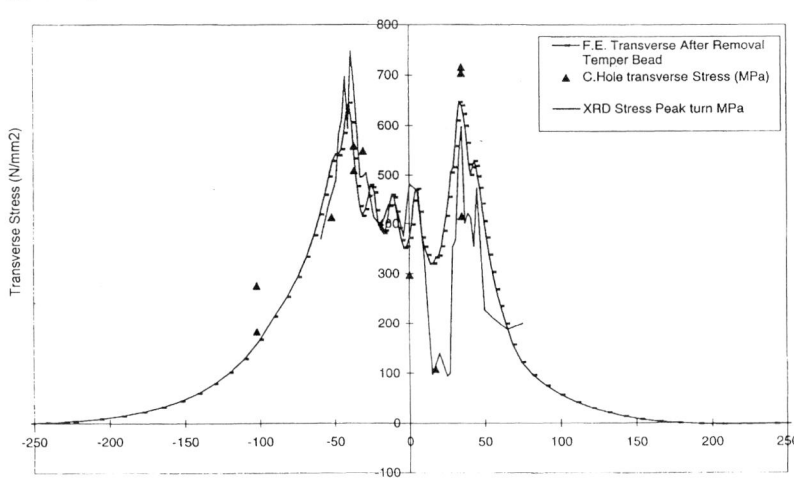

Figure 8: Comparison of measured and predicted residual stress distributions across the surface of the repair weld

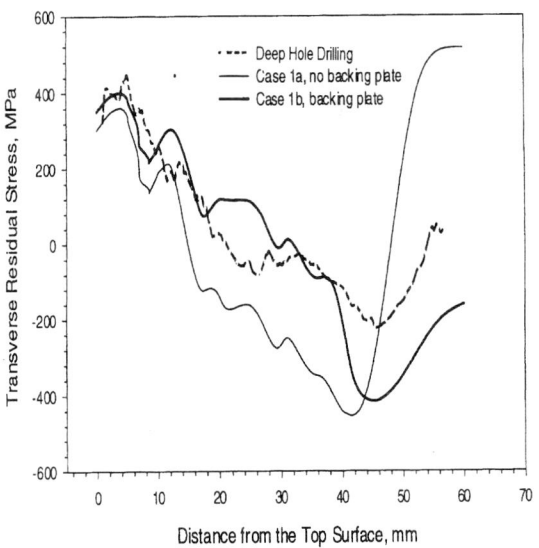

Figure 9: Comparison of measured and predicted residual stress through the thickness of the repair

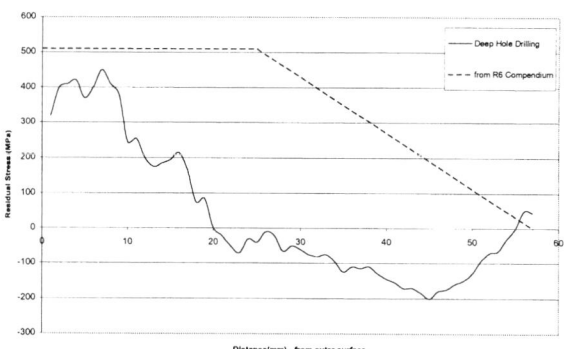

Figure 10: Comparison of R6 compendium and measured residual stresses in the repair

S690/003/99 *Sizewell A Power Station Boiler Repair*

S690/004/99

A regulatory view of the boiler shell weld repair

G B HEYS and **R E WATERS**
Health and Safety Executive, Nuclear Safety Directorate, HM Nuclear Installations Inspectorate, UK

ABSTRACT

NII's regulatory approach to the Sizewell A boiler shell safety case is described in the context of relevant UK legislation and guidance. The results of the periodic safety reviews of the boiler shell safety case, leading to the discovery of the defects, are summarised. Key boiler shell safety case issues are reviewed in relation to the Licensee's strategy for continued operation. The success of the UK's goal setting, non-prescriptive regime, as applied to the repair, is discussed. The paper considers the importance of project safety management arrangements, the use of modern nuclear design codes and redundant, diverse and validated (qualified) inspections to underwrite the repairs. The confidence gained from independent monitoring, audit, review and verification of the design, fabrication, inspection and tests, and safety case development is considered.

1. INTRODUCTION

The main legislation governing the safety of employees and the general public at nuclear installations is the Health and Safety at Work, etc. Act 1974 and the associated relevant statutory provisions which include the Nuclear Installations Act 1965 (as amended). The Nuclear Installations Act gives the Health and Safety Executive (HSE) the power to attach to the nuclear site licence such conditions as may appear necessary or desirable in the interests of safety. The periodic safety review of the boiler shell safety case identified the need for further inspection of the boiler shells at Sizewell A and the Nuclear Installations Inspectorate (NII) issued a Licence Instrument specifying these inspections. During these inspections, significant surface breaking cracks were discovered in three of the four boilers on Reactor 2.

The Licensee subsequently developed a strategy for continued operation. The regulatory approach to continued operation focused upon understanding the reasons for the cracking to ensure that any options for continued operation reduced risk as low as reasonably practicable. Key safety principles for the repair were agreed with the Licensee which centred on a nuclear code compliant repair supported by redundant, diverse and validated (qualified) inspections.

Considerable confidence was gained from independent monitoring, audit, review and verification of the design, fabrication, inspection and tests, and safety case development.

2.   REGULATORY REGIME

2.1 Primary legislation and regulatory framework
In the UK, legislation governing the safety of nuclear installations places the responsibility for the safe design, construction, manufacture, modification, commissioning and operation of nuclear installations with the licensee; that is with the station operator. UK safety legislation is non-prescriptive, NII does not prescribe detailed standards or codes of practice for nuclear plant. This responsibility rests with the licensee who is required to formulate the design safety criteria and standards, and the construction, commissioning and operating arrangements and procedures which will be used.

The duty of the NII is to see that appropriate standards are developed, achieved and maintained by the licensee, to ensure that the necessary safety precautions are taken and to regulate and monitor the safety of the plant by means of its powers under the licence.

As the boiler shell is a pressure vessel the Pressure Systems and Transportable Gas Containers Regulations 1989 are applicable. An Approved Code of Practice (ACOP) (1) was issued by the Health and Safety Commission in 1990. It is noted in the ACOP that a licensee, operating a pressure vessel at a nuclear licensed site, may need to provide additional evidence beyond the requirements of the regulations combined with the ACOP, to demonstrate that the systems are adequately safe for their nuclear safety duty.

The standard Conditions attached to the Nuclear Site Licence (2) are enforceable in the same way as regulations made under the Health and Safety at Work Act. In addition to its powers under the Nuclear Installations Act, NII can take enforcement action using the Health and Safety at Work Act. NII's powers under the site licence, and their use in regulation of the boiler shell weld repair, are described later in this paper.

2.2 Regulatory guidance
No human activity is entirely free from risk. The HSE publication "The Tolerability of Risk from Nuclear Power Station" (TOR) (3) presents HSE's position on the tolerable levels of individual and social risk to workers and the public from nuclear power stations. TOR also explains the ALARP (As Low As Reasonably Practicable) principle. NII regulates the industry to ensure that licensees meet the ALARP principle and relevant legislation. Application of the ALARP principle to the boiler shell repair is discussed at Section 4.1.

The NII has developed a set of Safety Assessment Principles (SAPs) (4) for guidance to inspectors. These were developed mainly for its own staff, but they indicate to the public what standards HSE, through NII, expects to see in designs for new nuclear plant. These principles set out the requirements for assessing the safety of designs. They reflect HSE's views given in TOR. The SAPs may also be used by the licensee to help formulate the safety case. NII looks for 'defence in depth' which includes minimising faults and the containment of hazardous material by using as many different complementary ways as are reasonably practicable.

NII sets safety goals and the aim is for licensees to set out how they will meet them. NII examines how these are met by the licensees and will take enforcement action where necessary. A balance has to be struck in how far NII should be involved in the design and assessment process. NII's safety assessment is a matter of careful selection (sampling) of what are the key safety issues and what aspects should be examined. NII does not involve itself in all aspects of the design but concentrates on the main issues and aspects of concern to safety. Licensees must carry out their own detailed assessment and audit of the design process from the safety point of view. NII seeks to satisfy itself that licensees have the organisation for this and that they are carrying out their functions effectively. The NII therefore intervenes rather than participates.

3. RESULTS OF PERIODIC REVIEWS OF BOILER SHELL SAFETY CASE

3.1 Purpose and objectives of periodic reviews

Throughout the design lifetime of the plant, and beyond, NII requires licensees to keep the plant and its performance under review to ensure that the safety case remains valid. The Magnox power reactors are the oldest now operating in the UK and as each approached the end of its original 'design' life, it was considered prudent that the licensee should carry out a comprehensive review of safety. These reviews became known as the Long Term Safety Reviews (LTSRs) and were to demonstrate that the plant could be operated safely beyond its design life. The main objectives were to:

- confirm that the plant was adequately safe for continued operation;

- identify and evaluate any factors that might limit safe operation of the plant in the foreseeable future; and

- assess the plant's safety standards and practices and introduce any improvements that were reasonably practicable.

The LTSR of the safety case for the boilers at Sizewell A identified the need for reasonably practicable ongoing inspections to support continued operation up to 31 March 1996, completing thirty years operation, when it was agreed that the first "Periodic Safety Review" would be carried out to support operation beyond thirty years.

As a requirement of the nuclear site licence, all nuclear licensees must carry out periodic reviews of their plant and the associated safety cases. For reactor plant, Periodic Safety Reviews (PSRs) are currently required every 10 years. The objectives for the PSRs are:

To extend the applicability of the existing safety case for a further 10 years and include worthwhile improvements in scope and reassessment of the extent of compliance with current safety standards and working practices.

- To make any necessary improvements which are desirable and reasonably practicable to plant and procedure.

- To carry out a systematic review of age-related degradation and ensure that suitable monitoring and surveillance schemes are in place.

- To produce an updated safety case which reflects the current plant configuration.

3.2 Results of periodic reviews of Sizewell A boiler shell safety case

NII's assessment of the Licensee's safety case presented for the boilers at Sizewell A for the PSR and in response to NII LTSR findings, identified a number of areas of uncertainty. This resulted in a requirement for additional inspection to support continued operation up to and beyond thirty years. The NII requirement for inspection of a significant proportion of seam welds was reinforced by uncertainties in the following areas:

Quality of original manufacture

- One of the boilers (2A, subsequently renamed 2C after refurbishment) failed the pre-service hydrotest (5). Uncracked plates were reworked and used to fabricate boiler 2C.

- Pre-service volumetric inspection employing radiography was undertaken after welding. No inspection was undertaken after Post Weld Heat Treatment (PWHT).

- High repair rates during boiler fabrication, in particular for refurbishment of boiler 2C following hydrotest failure, were reported in the construction records.

In-service inspection

- Volumetric in-service inspection of a limited length of seam welds revealed subsurface defects approaching the reference size assumed in the safety case for the PSR.

Fabrication

- A disproportionate number of pressure vessel failures have occurred in Mn-Cr-Mo-V, Ducol W30 type, steels. Lancaster (6) gave the view that particular care is required in welding and fabrication of Mn-Cr-Mo-V steels.

- The materials of construction are susceptible to hydrogen cracking and stress relief, or reheat cracking.

3.3 Inspection of boilers

In December 1995, NII issued a Licence Instrument specifying the inspections to be carried out for continued operation of Reactors 1 and 2 following consultation with the Licensee. The proposed inspections covered the most limiting locations identified from the stress and fracture analysis, and a range of construction parameters and operating conditions. This work resulted in over 20% of the total weld length of all eight boilers being inspected. During the inspections of Reactor 2's boiler 2C in February 1996, structurally significant surface breaking cracks were discovered at Course 6 to 7 circumferential seam weld. The Licensee then extended the agreed inspection programme to cover all accessible axial and circumferential seam welds. This resulted in nearly 100% inspection of all boiler shell seam welds on Reactors 1 and 2. During these inspections, in August 1996, structurally significant surface breaking cracks were also found in boilers 2A and 2D on Reactor 2, at the Course 6 to

7 circumferential seam welds. Their detection and diagnosis is described by Professor Flewitt and Dr Exworthy elsewhere in these proceedings (7).

Although defects were identified by these inspections in Reactor 1's boilers, NII was satisfied that an adequate safety case was made to permit the return to power of Reactor 1 in September 1996 and the PSR was cleared to permit this. However, to ensure that failure remains extremely unlikely, the Reactor 1 boiler shell welds will be subject to a long term programme of inspection and monitoring which will be reviewed regularly by the Licensee and regulated by NII. Reactor 2 remained shutdown whilst the Licensee undertook a programme of work to address the safety issues associated with the cracks found in boilers 2A, 2C and 2D.

4. REGULATORY APPROACH TO CONTINUED OPERATION

4.1 Demonstration that options for continued operation were ALARP
The PSR demonstrated that in the event of boiler shell failure, the reactors can be safely shut down and cooled, and that a large release of radioactivity was unlikely. However, the Licensee recognised the need to demonstrate that failure of boilers would be an extremely unlikely event and required a safety case which could claim high reliability.

NII required that any option for continued operation, in addition to being demonstrably safe to show that the fundamental principle of ALARP was observed, and to assure that safety margins were maximised and hence risk reduced. NII has developed guidance for internal use on the application of ALARP. Essentially there are two legs to demonstration of ALARP: these may be described as qualitative and quantitative. The quantitative demonstration is primarily a cost benefit analysis. For the Sizewell A boiler shell safety case, as it was impracticable to assign a failure frequency with a high degree of confidence, the qualitative demonstration was essential. NII's hierarchy of requirements for the Sizewell A boilers to demonstrate ALARP qualitatively were:

- Adherence to any relevant Approved Codes of Practice (ACOP) issued by the Health and Safety Commission or the Executive.

- Regulatory guidance - NII Safety Assessment Principles.

- Relevant modern codes and standards.

- Industry sector guidance - relevant good practice.

- Precedent.

Where relevant good practice is not clearly established, the licensee must assess the risk and determine the extent of measures to avert the risk. In coming to a judgement on the adequacy of the ALARP aspects of the options for continued operation, NII asked the Licensee to provide the following information:

- Results of detailed analyses undertaken to support options under consideration.

- A consideration of the strengths and weaknesses of each option, a judgement on likely success and any decisions made regarding their viability.

- A comparison with modern standards.

- A comparison with approaches, industry sector guidance, and best practice, adopted in the nuclear and conventional industries.

- Comparison with the requirements of Pressure Systems Regulations and ACOP.

- Estimates of radiation dose for the work.

4.2 Cause of boiler shell cracking

Before coming to a regulatory position on continued operation, a detailed review of the cause of cracking of the boilers was considered essential, both to reduce uncertainties and improve fundamental understanding of the root cause. Therefore the following information was sought from the Licensee:

- Boiler shell manufacturing records to determine any differences between cracked and uncracked welds.

- Boiler shell operating history and events to investigate any contribution from in-service growth.

- Assurance that the material properties assumed for the boiler shell were adequately conservative and bounded sound material adjacent to defective welds.

- The extent and degree of grain boundary cavitation damage ahead of the crack tip and the effect on mechanical properties.

- Thermal ageing.

- Results of boiler shell replication and boat sampling to establish the contribution of microstructure to the mechanism and mechanical properties.

- The effect of welding and PWHT processes on the propensity to stress relief cracking and their effect on microstructure and mechanical properties.

- Reconsideration of the reason for the pre-service hydrotest failure of boiler 2C.

All of the above contribute to an improved understanding of the root cause of cracking of boilers 2A, 2C and 2D, and hence why other boiler shell welds, fabricated by nominally the same process and materials, were not cracked. Whilst the Licensee went some way to adequately answering many of the issues listed above, there was still no definitive explanation of the root cause of cracking.

The existence of a defect that is metallurgical in nature (hydrogen or stress relief cracking) may be an indicator of poor mechanical properties. As a result, it was judged that there remained significant uncertainties in materials properties. Additionally there was lack of validated data, using ultrasonic methods, to demonstrate that defect sizing was conservative for creep cavitated material. As mechanical properties of cavitated material were not available, it placed emphasis on the need to ensure that defect sizes determined by non-destructive testing adequately bounded any creep damaged material. In May 1997 NII therefore judged that the only sustainable way to provide adequate answers to these uncertainties was by extensive and detailed intrusive boiler shell sampling and testing.

4.3 Regulatory approach to continued operation

At an early stage of the development of the Licensee's strategy to return Reactor 2 boilers to service, the Licensee was informed of our proposed regulatory approach. The safety case required a very high degree of assurance and so assessment would be guided by the Special Case Procedure (4), SAPs P70 and P71. This requires a comprehensive examination of all the relevant scientific and technical issues, taking account as appropriate of precedents set under comparable circumstances in the past. Additionally, an independent assessment should be carried out in addition to the checking provided as part of the design process. The objective of the Special Case Procedure is to confirm the adequacy of the design specification and that the manufacture, construction and commissioning satisfies that specification. It was decided that the principles used in the assessment of the design, construction and manufacture of Sizewell B Pressurised Water Reactor Incredibility of Failure components should be used as the precedent. This was used to guide NII's approach throughout the project.

While the Licensee was developing a strategy for continued operation, in May 1996, NII sought independent advice from experts in the field to review state-of-the-art of weld repair techniques for Mn-Cr-Mo-V steels (BS 1501-271) in operating plant to inform our judgement. The review covered the following issues: material development and application; pressure vessel failures; fabrication procedures, qualification and selection of consumables; codes and standards; non-destructive testing (NDT); material properties; integrity assessments; and service performance. The review clearly identified that a Post Weld Heat Treated (PWHT) repair was the option most likely to be successful and whose integrity could be demonstrated. However, the review confirmed the NII view, that care would be needed in design, selection of consumables and welding procedure, control of welding, PWHT, NDT and its validation. Consequently the Licensee was advised that a PWHT weld repair should be an acceptable option for continued operation provided it was properly conceived and implemented.

Whilst the issues discussed above were pursued by communication between NII and the Licensee, both by correspondence and meetings, the Licensee was evaluating and developing weld repair procedures. Once these were sufficiently well developed, the Licensee decided to evaluate the economic and technical feasibility of undertaking a post weld heat treated weld repair of the cracks. The Licensee wished to establish the regulatory related risks before embarking on an expensive project and dialogue with NII commenced to minimise these risks. NII informed the Licensee of the likely principles and criteria against which the safety case would be judged and these are discussed at Section 5. This communication and exchange of views was important in establishing safety acceptance criteria which both parties could agree.

5.   KEY SAFETY PRINCIPLES FOR THE BOILER REPAIR

5.1 Principles and criteria

Before commencement of repair activities the Licensee presented a Paper of Principle that defined the safety principles and criteria against which the repair would be undertaken. The Licensee sought NII's agreement to the Paper of Principle, through the arrangements made under the Licence. The aim of these safety principles and criteria was to ensure that the repair was developed and implemented to achieve and demonstrate adequately high integrity to assure safety for future operation.

The key safety principles against which NII judged the adequacy of the repair were:

- Adaptation of a relevant nuclear code (the preferred code being ASME Boiler and Pressure Vessel Code), any non-compliances or deviations from code to be adequately justified.

- Clear and auditable project safety management arrangements.

- Independent monitoring, audit, review and verification of the design, fabrication, qualification, inspection and tests, and safety case development.

- Redundant, diverse and qualified (validated) inspections to the highest standard, repeated at appropriate stages of the repair. Independent qualification for the inspections deemed to be most safety critical.

- Independent peer reviews.

- Boiler shell materials sampling and a materials testing programme to address relevant safety issues for the weld repair and return to service safety case.

- Optimisation of repair parameters and processes to prevent recracking.

- Avoidance or mitigation of damage to primary circuit and boiler components.

For the defects located at Course 6/7 seam weld, repair was possible. However, the defect located at circumferential seam weld between Courses 5 and 6 of Boiler 2C was removed by machining and hand dressing to a profiled excavation. This aspect of the repair is discussed further below.

5.2 Design codes and standards

To achieve compliance with NII SAPs, the design should be conservative and follow appropriate national or international codes and standards. The benefits of working to a recognised nuclear code are that the best practicable standards of manufacture, construction, quality assurance and inspection are embodied in such a Code and its use should help achieve adequate procedure validation and qualification. The Licensee chose to adopt the ASME Boiler and Pressure Vessel Code, Sections III and XI, and justify any non-compliances or deviations from code in the ASME Design Specification. Whilst non-compliances were inevitable, given the material of construction, it was judged that benefits could be derived from use of the ASME Code because of its previous use for many years in the nuclear industry world-wide, not only for construction and manufacture, but for in-service inspection and repairs. Therefore NII agreed with the Licensee's decision to adopt the ASME Code.

5.3 Project arrangements

A number of key requirements flowed from adaptation of the ASME Code. The Boilers were designated as Class 1 components. The requirements for project management and quality assurance for ASME Class 1 components are similar to the legal requirements of NII's standard Licence Conditions. The Licensee developed Project Implementation Arrangements that defined for example, organisation, responsibilities and interfaces. The arrangements confirmed the appointment of an Independent Third Party Inspection Agent (ITPIA), performed by Kennedy & Donkin, whose role encompassed that of the Authorised Nuclear (In-service) Inspector defined in the ASME Code. This also confirmed the appointment of Royal and Sun Alliance as Independent Design Appraisal (IDA) contractor, whose primary role was to certify the design reports.

A Quality Assurance programme and Design Specification were developed. The Design Specification was effective in defining the design requirements and applicable codes and standards to be used for the repair. The interface arrangements with the competent person under the Pressure Systems Regulations were also clarified.

The main objective was that the repairs should be controlled and implemented in a clear and auditable manner, and that adequate arrangements should be developed for this. Particularly confidence of correct implementation of irreversible steps was necessary as few regulatory hold points were specified. This was achieved in practice.

Before repair activities commenced, the Licensee was required to develop and implement arrangements for the reporting, assessment, and sentencing of defects revealed during repeat NDT at each stage of the repair. This was achieved through the appointment of a Project Defect Assessment Panel which was responsible for reviewing inspection findings, considering, sentencing and agreeing any remedial actions. The IDA and peer reviewers were represented on the panel and provided the strong independent element that the NII considered so important. Defect acceptance criteria and a weld repair rectification procedure were developed and justified by the Licensee.

5.4 Independent reviews

Considerable importance was attached to the appointment of a number of bodies to undertake independent monitoring, audit, review, verification and certification of the design, fabrication, qualification, inspection and tests, and safety case development. This need reflected the nature of the significant technical challenges of the project, irreversible steps and consequences of boiler shell failure. These functions were fulfilled by the ITPIA, the IDA and the Independent Welding Approval (IWA). The IWA acted to ensure the welding procedures and personnel were approved using the Safety Assessment Federation (SAFed) guidelines (8).

The Project Implementation Arrangements document appointed Independent Peer Reviewers with responsibility to undertake rolling peer reviews in their relevant areas of expertise. This commitment to undertake expert peer reviews was welcomed, and is believed to have brought significant benefits to the project.

5.5 Optimisation of repair processes

In considering the Licensee's request for NII's consent to commence welding on boiler 2D, work undertaken to develop and optimise the repair processes to prevent recracking was judged to be of considerable significance. The work also contributed further to the understanding of the root cause of cracking of the boilers. Confidence was gained from the following work:

- Development and optimisation of welding process to produce a fine grained HAZ. Stress relief cracking in boiler shells 2A and 2D occurred in coarse grained HAZ (7).

- Optimisation of weld preheat to minimise risk of hydrogen cracking. Hydrogen cracking was a likely contributor to cracking in boiler shells 2A, 2C and 2D (7).

- Increasing the preheat to 200°C to off-set the low heat input required by the temper bead welding procedure and minimise the likelihood of hydrogen cracking.

- Run off plates tack welded to the boiler surface along the outer edges of the weld preparation to ensure adequate grain refinement of the HAZ close to the boiler surface.

- Proceeding directly from welding to PWHT, without the loss of preheat to avoid the combination of high residual stresses and ambient temperatures.

- Controlled ramp rates to PWHT soak temperature to minimise dwell times in creep ductility trough, and minimisation of thermal gradients to reduce thermal stress (7, 9, 10).

- Optimisation of heated bandwidth to minimise thermal stress (9). Professor Burdekin (11) considers that the limited heated bandwidth, used for local heat treatments (12) during original manufacture of the Sizewell boilers, aggravated any tendency for stress relief cracking.

- Uniform temperature gradient away from the peak to minimise thermal stress (9).

- Optimisation of PWHT soak temperature to maximise creep ductility and minimise creep crack growth rate at PWHT temperature (9).

- Measures taken to control and monitor the PWHT and welding.

5.6 Inspection issues due to absence of hydrotest

At an early stage in developing the repair procedure the Licensee undertook a review of the feasibility of a pressure test. Whilst a hydraulic or pneumatic pressure test was feasible in principle such tests posed major engineering difficulties and would lead to contaminated waste water disposal problems. NII accepted that the Licensee's safety case demonstrated that a pressure test was not practicable.

The Licensee therefore appealed to ASME Code Case N-416-1 (13), which gives recommendations in lieu of performing a system hydrostatic pressure test for welded repairs, applicable to Class 1 components Section XI, Division 1. An ALARP justification was required for the absence of an hydrotest and discussion of the merits in relation to the repair activities. This guided and significantly influenced the approach and design specification for the profiled excavation of the defect at Course 5/6 in boiler 2C.

As a hydrotest was not practicable, the final proof of integrity lay solely with inspection. Therefore NII needed to be confident in the sufficiency and adequacy of these inspections. The Licensee provided this confidence by validating the non-destructive examinations to the highest standard defined by the European Network for Inspection Qualification (ENIQ) (14). The appointment of an Independent Qualification Body, (with a role similar to that for the Inspection Validation Centre for Sizewell B) as defined in ENIQ, was a key requirement. Redundant, diverse and qualified (validated), in-process and pre-service inspections to the highest standard, repeated at appropriate stages of the repair was of paramount importance to a successful regulatory outcome to the repair.

5.7 In-process inspection
Given all the measures to prevent a recurrence of the cracking which occurred during boiler construction, if stress relief cracking occurred during the repair this would have cast severe doubts on the adequacy of the safety case and led to a full review of the whole project before considering the next steps.

As a result it was judged important to be able to distinguish between any stress relief cracking and other planar defects which might occur in the repaired weld, particularly at the HAZ of the repair weld and in the HAZ of the original construction weld. The Licensee was required to do all that was reasonably practicable to distinguish between stress relief cracks and other planar defects and establish the amount of growth of any defects which may occur during the PWHT to a high level of reproduceability.

5.8 Materials sampling and testing
A key NII Requirement arising from the Periodic Safety Review was for the Licensee to complete a comprehensive materials testing programme to complement and build upon the available database for the materials of construction of the Sizewell A boilers. In advance of agreeing to the Paper of Principle for the repair, a major extension to the programme to cover boiler shell sampling and a materials testing programme, to address relevant safety issues for the weld repair and return to service safety case was defined.

As discussed at Section 4.2, there was a clear regulatory need for intrusive materials sampling of the boiler shells to reduce uncertainties primarily in materials properties, but also in NDT sizing capability. This was needed to be confident that any proposed repairs had a high chance of success, and all that was reasonably practicable would be done before agreeing to an irreversible process which could foreclose options. NII had to be satisfied that this was carefully controlled to prevent the repair being extended. Boiler shell sampling was effective in providing detailed information on:

- the extent and mechanism of cracking;

- the contribution of microstructure to the cracking;

- mechanical properties of the ligament below the repair weld, and the integrity of the remaining ligament for the profiled excavation of Course 5/6 defect;

- improved confidence in ultrasonic inspection defect sizing capability for creep cavitated material; and

- ensuring removal of defective material thus increasing the chance of effecting a successful repair.

This work also resulted in the Licensee deciding to undertake a fully circumferential repair due to virtually fully circumferential creep cavitated material being detected close to the external surface.

The key objectives that NII considered the materials testing programme needed to address were:

- To identify the optimum PWHT temperature. The objective here was to ensure that creep ductility was adequate to accommodate relaxation of residual and thermal stress without causing further stress relief cracking but without unduly softening the material by over-tempering.

- To resolve generic concerns regarding the toughness behaviour of coarse grained HAZ.

- To support the weld procedure qualification tests.

- To improve the database of materials property data for integrity assessments of Sizewell A boiler materials.

- To demonstrate that the weld procedure tests and mechanical property programmes were carried out on materials representative of the boiler shell plates and welds.

The Licensee satisfied the NII that the materials test programmes comprehensively covered the range of materials and conditions to be expected after completion of repair activities. Certain longer term tests are ongoing, such as creep testing, to complete understanding of materials properties and to provide further assurance that the properties are as expected.

5.9 Protection of pressure circuit components

A safety issue of paramount importance was that the repair activities did not prejudice the integrity of the primary circuit pressure boundary, in particular the Reactor Pressure Vessel and boiler internal components. In advance of welding and PWHT, the Licensee undertook detailed analyses of the effect of the repairs on components and the pressure boundary potentially affected by the repair. Key principles were identified by NII's Agreement to the Paper of Principle and subsequently detailed arguments presented in Stage Submissions. The approach included: temporary engineering modifications; extensive monitoring; administrative controls during welding and PWHT; and engineered measures for mitigation of adverse indications. This was supported and confirmed by: inspection of a comprehensive range of components pre and post PWHT; analysis of results from monitoring data to confirm the pre-repair stress analyses; and pre-service testing of components. The Licensee demonstrated that the extent of testing, inspection and analysis was commensurate with the safety categorisation of the components and the consequences of their failure.

5.10 Profiled excavation of Course 5/6 defect

For the defect located at the Course 5/6 circumferential seam weld, the Licensee demonstrated that a weld repair was not reasonably practicable due to restricted access to this region of the boiler internal surface. As a result, it was deemed impossible to install insulation at this location to protect boiler internal structures from heat during welding and PWHT without removal of the boiler internals. After due consideration of alternative approaches, the Licensee decided to undertake a profiled excavation of the defect. NII required that the proposed design was optimised to ensure adequate coverage for defect detection and sizing using qualified volumetric inspection procedures. The key principles agreed with the Licensee to ensure an adequate profiled excavation are listed below:

- Maintain minimum design wall thickness.

- Ensure that the corners of the profile are away from the weld HAZ.

- Optimise the slope of the profiled excavation to minimise peak stress.

- Peak stress to be less than yield and minimised.

- Fillet radii to be optimised to minimise peak stress increment.

- Defect to be blended circumferentially.

- Machining (using jig based fixtures) and surface conditioning of the profiled excavation.

- Need for Code compliant design analysis.

- Need for limit to be set, before commencement of machining, on maximum depth of excavation.

- Integrity assessment of defect tolerance.

The safety case for return to service clearly demonstrated achievement of the above criteria and demonstrated the defect tolerance of the design.

6.   REGULATORY APPROACH TO THE BOILER REPAIR

6.1 Regulatory control

The Conditions attached to the Nuclear Site Licence require the classification of modifications according to their safety significance and where appropriate division of the modification into stages. Where NII specifies the licensee must not commence, nor proceed, from one stage to the next without NII's consent. Adequate documentation to justify the safety of the proposed modification is required and where appropriate be provided to NII.

For the boiler shell repairs, which were Category 1 plant modifications, NII agreed a series of hold points with the Licensee before the Paper of Principle was submitted. This consisted of effectively four hold points which are described below.

Hold Point 1 (essentially the start of the project for regulatory purposes) required NII's Agreement to the Licensee's Paper of Principle for the Boiler shell repairs. Agreement was given on 27 October 1997. This gave NII's Agreement to:

- The proposed programme of development and implementation of a weld repair procedure and post weld heat treatment for the defective Course 6/7 circumferential seam welds in boiler shells 2A, 2C and 2D.

- Profiling of the Course 5/6 defect in boiler 2C.

- The development of a safety case for the return to service of Reactor 2.

- The project arrangements.

Hold Points 2(2D) and 2(2A/2C) were commencement of welding on boiler 2D and subsequently boilers 2A and 2C. Consent to commence welding of boiler 2D was granted by NII on 14 August 1998 and boilers 2A and 2C on 25 October 1998.

The regulatory route for return to service of Reactor 2 was discussed and agreed with the Licensee. Commissioning was in two stages. The first involved a leak test of the circuits, confirmation of monitoring equipment performance, and confirmation of the reinstatement of ancillary pipework and circuits. This was carried out using non-nuclear heating. The second stage was commissioning with nuclear heating.

The third Hold Point was before first pressurisation to 150 psig at 160°C and to begin commissioning; Consent to pressurise was issued by NII on 3 March 1999.

Hold Point 4 was start of nuclear heating. The Licensee provided evidence of satisfactory completion of stage 1 commissioning. Regulatory control of nuclear heating was provided by Consent to start-up the reactor after a statutory outage. Consent was given by NII on 24 March 1999.

The next stage in the project was completion of commissioning (stage 2) and return to routine operation. Following return to power and within three months of Reactor 2 start-up, the Licensee carried out volumetric and surface NDT to provide further assurance of the integrity of the repairs. This was satisfactorily completed in May 1999. A further check on the quality of the repairs will take place in Summer 2000 when Reactor 2 shuts down for its next statutory outage. During this outage the Licensee will repeat the full sequence of non-destructive testing of the repairs that took place after PWHT. Assuming these tests are successful it will mark the end of the project and return to routine operation.

To ensure that failure remains extremely unlikely, the Reactor 2 boiler shell welds will be subject to a long term programme of inspection and monitoring which will be reviewed regularly by the Licensee and regulated by NII.

6.2 NII Project Team Approach

Due to the potential risks posed to workers and the public arising from an inadequately conceived and implemented repair, NII assigned considerable effort to ensuring the safety of the completed repair. A small NII project team was established to regulate the return to power operation of Reactor 2. Expertise was drawn upon from within NII to assess and inspect safety case development and boiler repair implementation at site. Specialists in the following areas provided input to the regulatory process: welding and fabrication; heat treatment; materials; stress and fracture analysis; design codes and standards; non-destructive testing; quality assurance; safety management; conventional safety; and project management. The project team, comprising the authors and a project inspector, decided at the outset to take a proactive approach to regulation of the project. This was considered necessary due to the size and complexity, both managerially and technically. A strategy was developed, the primary objectives were:

- To provide reassurance regarding those items which could potentially present an increased risk to the public and workforce if poorly conceived or executed.

- To provide NII with the necessary confidence, from sampling, that the Licensee's process was working.

- To provide additional confidence in the safety case.

- To ensure that NII maintained the necessary independence from the Licensee's process.

- To minimise the number of hold points to that which the NII considered necessary for its regulatory purposes.

- To provide the basis for sound regulatory decisions at hold points.

- To reduce unnecessary delays at hold points.

A programme of work was developed identifying the Licensee's activities to inspect and assess. Well in advance of each hold point, a list of safety related documentation was identified. This included reporting completion of commitments from the prior stage of the project and safety documentation justifying progression to the following stage. Key documentation at each hold point included reports from the independent bodies and peer reviewers and their recommendation to proceed to the next stage. This approach was effective in achieving the above objectives and ensured that the team kept abreast of developments as they occurred.

From commencement of the boiler repair project to Consent to Hold Point 4, a total of 40 meetings were held with the Licensee to monitor and review progress. These meetings included: technical review meetings where technical progress was reviewed; regulatory project progress meetings to discuss safety and regulatory issues; NDT qualification progress meetings where qualification issues were discussed; and meetings with independent bodies. Twenty three site inspection visits were completed, lasting typically 3 days, and on occasion team inspections. Team members also visited Mitsui-Babcock to inspect arrangements for welder qualification and requalification. Whilst significant NII resource was expended on the project, considerable importance was attached to, and benefit derived from, the work of the independent bodies and the peer reviewers.

Whilst a number of technical issues arose during the course of the project, all of these were satisfactorily addressed. Possibly the most technically challenging area was the NDT because of the absence of a hydrotest, and of its potential to delay the project due to the qualification process. This is reviewed by Baborovsky (15) elsewhere in these proceedings.

6.3 Operator radiation dose
Radiation dose was managed by the Licensee throughout the project and the NII was satisfied with the approach. An accredited health physicist attended all the morning meetings of the project and therefore had an input to the daily management of work. Taking account of the amount of work in and around the primary circuit, the committed doses were low. The project may not have been viable had this not been the case. The total committed dose for the project, over its 15 month duration to 28 February 1999, was reported by the Licensee as 132.5 man mSv, and the highest dose to an individual, who was a contractor, was 5.5 mSv. Most of the dose was associated with the boiler entries.

6.4 Conventional Safety
There were no Lost Time Accidents. The NII was satisfied that accidents had been reported and that generally, these were minor. Working relations between contractors and the Licensee appeared to be good on the evidence of frequent site inspections. It is considered that this provides additional confidence in the quality of the work, as it shows appropriate control and supervision of work and management of safety.

7. CONCLUSIONS

7.1 The success of the legislation governing the safety of employees and the general public at nuclear installations and in particular the Nuclear Site Licence in regulating the repair has been described. The paper clearly demonstrates the effectiveness of the UK goal setting, non-prescriptive, regulatory regime which provides a flexible framework to exercise regulatory control of a major project such as the Sizewell A boiler shell repair.

7.2 The periodic safety review of the boiler shell safety case identified the need for further inspection of the boiler shells at Sizewell A and the NII issued a Licence Instrument specifying these inspections. During these inspections, significant surface breaking cracks were discovered in three of the four boilers on Reactor 2.

7.3 Achievement and demonstration of the following safety criteria and principles allowed NII to grant Consent for continued operation of Reactor 2, following completion of the post weld heat treated weld repairs to the defective boilers.

- Adaptation of a relevant nuclear code, and non-compliances or deviations adequately justified.

- Clear and auditable project safety management arrangements.

- Independent monitoring, audit, review and verification of the design, fabrication, qualification, inspection and tests, and safety case development.

- Redundant, diverse and qualified (validated) inspections to the highest standard, repeated at appropriate stages of the repair. Independent qualification for the inspections deemed to be most safety critical.

- Boiler shell materials sampling and a materials testing programme to address relevant safety issues for the weld repair and return to service safety case.

- Optimisation of repair parameters and processes to prevent recracking.

- Avoidance or mitigation of damage to primary circuit and boiler components.

7.4 Considerable confidence was gained from independent monitoring, audit, peer review and verification of the design, fabrication, inspection and tests, and safety case development.

8. ACKNOWLEDGEMENT

The views expressed in this paper are those of the authors and do not necessarily represent the views of the Inspectorate. The authors wish to acknowledge the contributions from many in NII, in the Licensee's organisation and in the various contractors involved in the project whose work resulted in a successful regulatory outcome to the boiler repair project.

9. REFERENCES

1.   Health and Safety Commission, (1990). "Safety of pressure systems. Pressure Systems and Transportable Gas Containers Regulations 1989. Approved Code of Practice." HMSO. ISBN 0 11 885514 X.

2.   Health and Safety Executive, (1994). "Nuclear Site Licences under the Nuclear Installations Act 1965 (as amended). Notes for Applicants". London, HMSO. ISBN 0 7176 0795 0.

3.   Health and Safety Executive, (1992). "The Tolerability of Risk from Nuclear Power Stations". London, HMSO. ISBN 0 11 886368 1.

4.   Health and Safety Executive, (1992). "Safety Assessment Principles for Nuclear Plants". HMSO. ISBN 0 11 882043 5.

5.   West of Scotland Iron and Steel Institute. "Special Report on Failure of a Boiler During Hydrostatic Test at Sizewell Nuclear Power Station." 1964.

6.   Lancaster, J. (1992). "Handbook of Structural Welding". Cambridge, Abington Publishing.

7.   L F Exworthy, B J C Ellis and P E J Flewitt. Proceedings of I. Mech. E. Seminar (1999). "The Boiler Shell Weld Repair at Sizewell A Nuclear Power Station". Paper 2.

"Metallurgical evaluation of the nature and origins of the cracking and the implications for the repair strategy".

8.   Safety Assessment Federation Report (1998), ISBN 1901212254.   "Welding Procedures and Welders Guidelines on Approval Testing."

9.   E J McDonald, N Hunter and W Bell. Proceedings of I. Mech. E. Seminar (1999). "The Boiler Shell Weld Repair at Sizewell A Nuclear Power Station".   Paper 6. "Specification, development and optimisation of the welding and post weld heat treatment procedure"..

10.   A N R Hunter, E J McDonald , R Moskovic and M Lamb. Proceedings of I. Mech. E. Seminar (1999). "The Boiler Shell Weld Repair at Sizewell A Nuclear Power Station". Paper 8. "Materials challenges".

11.   F M Burdekin (1997). Private communication.

12.   F M Burdekin. British Welding Journal, September 1963. "Local stress relief of circumferential butt welds in cylinders."

13.   ASME Code Case N-416-1. "Cases of ASME Boiler and Pressure Vessel Code". 15 February 1994. "Alternative pressure test requirement for welded repairs or installation of replacement items by welding, Class 1, 2 and 3. Section XI, Division 1."

14.   ENIQ Report No. 2. Reference No. EUR 17299 (1997). "European Network for Inspection Qualification. European Methodology for Qualification (2nd Issue)".

15.   V M Baborovsky. Proceedings of I. Mech. E. Seminar (1999). "The Boiler Shell Weld Repair at Sizewell A Nuclear Power Station". Paper 11. "NDT Qualification".

S690/005/99

Underwriting the integrity of the repair pressure circuits

P J JEANS, H G MUNRO, and E G TAYLOR
BNFL Magnox Generation, Berkeley, UK

ABSTRACT

It was recognised at the outset that practical difficulties would prevent an overpressure test being undertaken following completion of the weld repairs at Sizewell A Power Station. In addition to general compliance with code requirements the integrity of the repaired boiler shells was therefore underwritten by rigorous quality control procedures for welding, post weld heat treatment (PWHT) and profiling and by extensive and qualified inspections of the weld repairs. These measures were supported by both design and defect tolerance assessments of the welds.

Precautions, including separating the reactor gas outlet duct and off-loading the superheater tube bank support brackets, were taken to ensure that the rest of the pressure circuit was adequately protected during PWHT. Confirmation of the adequacy of this protection was provided by careful monitoring of temperatures and displacements during PWHT and by extensive inspections undertaken before and after PWHT on all potentially affected plant.

1. INTRODUCTION

There are four main gas circuits connected to each of the two reactors at Sizewell A Power Station, a typical circuit being shown in Figure 1. Figure 2 shows a general assembly of the boiler shells which are 27.9m high and have an internal diameter of 6.86m. The thickness of the shell is 57 mm. Each boiler shell was fabricated from BW87A low alloy plate steel and consists of seven cylindrical courses, each made up of three plates, and two dome ends. The bottom dome is made up of six petal plates whilst the top dome comprises two sections of four and six plates.

With respect to the boiler shell repairs the principal nuclear safety concern was the possibility of failure of the boiler shell in service with subsequent failure to achieve satisfactory trip, shutdown and post trip cooling. The boiler shell safety case was therefore based on both the improbability of shell failure and the acceptable radiological consequences of such a failure. The repairs did not introduce any substantive changes in relation to the issues of trip, shutdown and post trip cooling. The prime requirement was therefore to demonstrate an acceptable level of boiler shell integrity under all service conditions. Although it was required to show that boiler shell failure is an infrequent event, defined as $<10^{-3}$ per reactor year, it was intended that a much higher level of integrity would, in fact, be demonstrated by considering quality of manufacture, functional testing, structural analysis and forewarning of failure aspects.

In order to demonstrate a high level of integrity it was decided to remove the major defects in the Course 6/7 welds completely by excavating the welds over the full length of the defects and to carry out a weld repair and post weld heat treatment (PWHT) on the resulting excavations.

The Course 5/6 defect in Boiler 2C was significantly shallower and less extensive than the Course 6/7 defects. Furthermore, compared with the Course 6/7 weld, there were substantial additional practical difficulties associated with the PWHT of the Course 5/6 weld. In particular considerable difficulties had been identified associated with the installation of internal insulating materials in this region. However comprehensive insulation coverage would be required to avoid the risk of high localised shell stresses and to avoid exposing boiler internal components to excessive temperatures. To overcome these difficulties would have resulted in disproportionate costs in relation to the benefits derived from a weld repair. It was therefore decided to remove the Course 5/6 defect by excavating to a specified profile, rather than to effect a weld repair in this case.

2.   PRINCIPLES AND CRITERIA

In order to underwrite the integrity of the repaired pressure circuits a set of principles and criteria were established at the outset. Compliance with these principles would then provide confidence that a satisfactory outcome would be achieved.

Noting that there was no direct ASME equivalent to the BW87A shell material it was nevertheless proposed that the weld repairs and profiling be undertaken in accordance with ASME code requirements (1) for a Class 1 component and that any non-compliances which arise would be justified. Assurance of the future operational integrity of the repaired and profiled welds would then be provided by meeting the stringent acceptance criteria of the code.

Although the general defect acceptance criteria specified are onerous and do not permit any crack-like indications, the code allows larger defects to be accepted by specific agreement. It was therefore proposed that the significance of any defects which occurred during the repair process and which exceeded the general criteria would be assessed on a fitness for purpose basis using the R6 assessment route (2), and only if appropriate would further remedial actions be proposed.

The specific Principles which were established at the outset and subsequently adopted in the development of the repair, are presented in 2.1 - 2.11 below.

2.1   The Course 5/6 defect in Boiler 2C was to be removed by blending to an appropriate profile using an approved and validated machining process to result in a profiled weld compliant with the requirements of ASME III for a Class 1 component and of adequate strength to sustain all operational and postulated fault loadings.

2.2   The major defects in the Course 6/7 welds of Boiler Shells 2A, 2C and 2D were to be removed completely and the resulting excavations repaired using a manual welding procedure which would be approved and validated in compliance with the requirements of ASME XI and resulting, after PWHT, in a repaired weld of adequate strength to sustain all operational and postulated fault loadings.

2.3   A well defined and adequately controlled PWHT, compliant with the requirements of ASME III, and capable of reducing residual welding stress levels and improving material properties, was to be undertaken on the Course 6/7 repaired welds on Boiler Shells 2A, 2C and 2D.

2.4   It was to be demonstrated that adequate precautions were taken to ensure that the thermal expansion of the boiler shell during the PWHT would not prejudice the integrity of either the reactor pressure vessel or gas ducts.

2.5   It was to be demonstrated that application of the PWHT would not adversely affect the integrity of boiler internal components such as tube banks and support beams.

2.6   It was to be demonstrated that adequate precautions were taken to ensure that the PWHT would not prejudice the integrity of either the boiler shell or any attachments during its implementation and that the future operational integrity of the boiler shells and attachments would not be prejudiced.

2.7   The remedial activities were to be undertaken in such a way as to ensure the availability of an adequate cooling capability for the shutdown reactor.

2.8   All proposals for the implementation of activities associated with the repairs and profiling were to be accompanied by detailed method statements and would be subject to rigorous QA controls.

2.9   Sufficient inspections were to be carried out, using approved and validated NDT procedures, to confirm that the repaired and profiled welds were of the required integrity to sustain all normal operational and postulated fault loadings.

2.10   An appropriately resourced recommissioning procedure was to be established for returning the repaired boilers to service.

2.11   Prior to return to service a programme of future inspections would be specified commensurate with the requirements of the safety case.

3.   POST REPAIR TESTING

The rigorously controlled application of the repair procedures together with comprehensive inspections and demanding acceptance criteria was expected to enable substantial fault pressure margins to be demonstrated for the repaired and profiled welds, and a robust safety case to be established for the return to service of Reactor 2 at full pressure, without conducting an overpressure test.

Nevertheless it was recognised at the outset that it would be a general requirement of ASME XI (1) to carry out a pressure test to a pressure in excess of normal operating or fault values following completion of the proposed repairs and prior to return to service. In order to minimise the consequential risks of such a test a hydrostatic pressure test would normally be employed.

However ASME XI Code Case N-416-1 indicated that exceptions to this general requirement had previously been permitted. It noted that, in lieu of performing the hydrostatic pressure test, a system leakage test may be used providing that inspection is carried out in accordance with the methods and acceptance criteria of ASME III and that prior to or immediately upon return to service a visual examination and system leakage test at normal operating pressure and temperature is carried out.

The contribution provided by an overpressure test to enhancing confidence in the operational integrity of a pressure vessel may be summarised as:-

1. It proves the design of the vessel and quality of fabrication.

2. It facilitates 'stress shakedown' which offsets high stresses arising from fabrication or geometrical features.

3. It provides a 'proof pressure test argument' for defects that may not be detected.

For the repaired and profiled welds under consideration here the simple weld and profile geometries were expected to give confidence in the design whilst rigorously controlled repair procedures, thorough inspections and well characterised material properties would provide confidence in the fabrication quality. The benefits of 'stress shakedown' were not relevant for the repair welds since pressure stresses would be low and residual stresses would be limited by PWHT.   For the profiled Course 5/6 weld care would be taken to ensure that the machining process did not introduce significant residual stresses and that the geometry of the profile did not lead to any significant stress concentration, thus the benefits of shakedown were not relevant here.   Compensation for the 'proof pressure test argument' would be provided by the rigorous inspections proposed and the substantial defect tolerance anticipated for both the repaired and profiled welds.   It was therefore concluded that confidence in operational integrity would not be significantly enhanced by an overpressure test.

Nevertheless a study was undertaken to establish the feasibility of carrying out such a test on the boiler shells. It would have involved filling the boiler shells with approximately 800 tonnes of clean demineralised water. The study indicated that, while feasible in principle, such a test would pose major engineering difficulties as well as contaminated water waste disposal problems. Post test drying of the boiler would also present considerable difficulties. However an overriding factor was the requirement within existing Station Operating Instructions, arising from material property considerations, to limit the internal pressure to 50psi at temperatures below 150 $^\circ$ C. In view of this it was considered essential that any pressure test would have to be undertaken at a minimum temperature of 150 $^\circ$ C. The consequential advantage of a hydrostatic test over a pneumatic test, i.e. relatively low stored energy levels, would be lost at temperatures over about 115 $^\circ$ C and hence it was concluded that a hydrostatic test was not practicable.

Although the practical difficulties associated with conducting a pneumatic pressure test would have been less onerous than those associated with a hydrostatic test they would have been likely to impose substantial time and cost penalties. It would be necessary to isolate the circuit, probably by using blanks at the main gas valves and, noting the requirement for temperatures in excess of 150 $^\circ$ C, an external heating source would be needed to achieve pressures in excess of 150 psig.

In view of the high degree of confidence provided by the application of rigorous QA procedures and code requirements during repair, the avoidance of significant stress concentrations, comprehensive post repair inspections using validated methods and stringent defect acceptance criteria it was concluded that the limited additional confidence which would have been provided by an overpressure test was not necessary. Noting the considerable associated practical difficulties, it was therefore concluded on an ALARP basis, that an overpressure test was neither required nor reasonably practicable.

4. PROFILING COURSE 5/6 DEFECTIVE WELD IN BOILER 2C

The defect in the Course 5/6 weld in Boiler 2C was removed by excavating, in stages, to a specified profile. Local excavations at three positions were initially undertaken to establish confidence in the required excavation depth. Material test samples were then removed from the weld and adjacent plate to augment the ongoing materials test programme. The final profile was achieved by excavation using a combination of machining with purpose designed milling equipment, and blending by hand grinding to pre-set profile gauges. The milling machinery was hydraulically powered and was mounted on a track parallel to the Course 5/6 weld seam. The track was attached to the boiler shell with mounting brackets fastened by studs welded to the shell. The milling equipment included design features to prevent excessive removal of material and provided facilities enabling accurate control to be readily achieved. Completion of the profile was achieved by hand grinding using a process which had been shown to introduce negligible residual stresses. The geometry of the final profile achieved was essentially a flat bottomed groove approximately 200 mm wide and 14 mm deep with side walls inclined at 30^0 to the base. The length of the excavation was approximately 1m and at its circumferential ends it was blended tangentially. The profile was shown to result in stress levels which comply with the requirements of ASME III for all normal operational and postulated fault conditions.

5.   IMPLEMENTATION OF WELD REPAIR

The major defects in the Course 6/7 welds of each of the boilers were removed by excavating the full circumference of the boiler shell. The excavations were formed by machining a flat-bottomed groove approximately 70mm in width and with sides tapered at an angle of 65 to the boiler shell surface. The depths of the excavations were generally about 25mm but they were locally increased in some areas, to a maximum of about 44mm, to ensure the removal of all significantly damaged material and all significant defects close to the surface of the weld preparation. Machining was carried out using the hydraulically powered milling equipment described in Section 4 above. Prior to repair welding the completed weld preparations were subjected to thorough inspections and metallographic examinations. The results of these inspections demonstrated that the excavations were in a suitable condition for repair welding.

The weld repairs were carried out on all three boiler shells using a manual metal arc process with a pre-heat of 200 $^{\circ}$C over an axial width of approximately 2m centred on the weld.

The essential parameters of the weld repair procedure, which complied with the intent of ASME III and ASME IX, were maintained within specified limits to ensure that the weld repair would have a fine grained HAZ with minimum risk of stress relief cracking. The welding was carried out in accordance with approved method statements with only minor adjustments to simplify implementation. Inspections undertaken following welding revealed only minor welding defects and no local repairs were required on any boiler.

6.   POST WELD HEAT TREATMENT

A local, but fully circumferential, PWHT was undertaken on each of the repair welds. The procedure was optimised to achieve good creep ductility and adequate strength based on a materials test programme. A nominally constant temperature in excess of 650 $^{\circ}$C (and with a target of 675 $^{\circ}$C $\pm$ 10°C) was maintained for a period of 2.5 hours over the full circumference of the boiler shell for an axial length of 400mm centred on the repaired weld. Outboard of this band, for a distance of 2m, a linear profile was maintained with temperatures reducing to 260 $^{\circ}$C at the limit of the controlled region. The heat treatment was fully compliant with the requirements of ASME III and BS5500 (3) and was supported by thermal stress analyses using Finite Element techniques. Care was taken to ensure rapid heating rates through the low ductility temperature range whilst ensuring compliance with all code requirements.

It was recognised that implementation of the PWHT would result in a thermal expansion of the boiler shell and, in order to eliminate any possibility of damage to the gas duct or RPV outlet nozzle during this process, the reactor outlet duct was separated at the bolted joint above the main gas valve. A restraint system was erected around the flange to enable separation to be effected in a controlled manner with measurement of the displacements of the separated flanges, changes in the forces of the counterbalance hanger units and the strains in the duct wall at the boiler inlet nozzle. No unexpected behaviour was observed during separation, PWHT or reinstatement.

The superheater tube bank support brackets are fitted in Course 6 at a minimum distance of about 1.0m from the Course 6/7 weld and were expected to be subjected to a maximum temperature of about 520 $^{\circ}$C during PWHT. Thus, as a precaution, and to minimise the risk of damage as a result of the PWHT, the superheater tube bank support brackets were deloaded, prior to PWHT, by suspending the tube banks from a temporary support structure attached to the inlet cone.

To gain access for installing internal insulation it was necessary to remove both units of the gas bypass valve located above the superheater tube bank and all of the associated bypass tubes. It was also necessary to remove the transverse baffles from each side of the bypass tubes and baffles around the vertical manway.

For all boilers the PWHT was carried out in accordance with approved method statements and quality plans and the intended heating rate, temperature distribution and soak temperature were achieved. There were minor instances of temperature excursions outside allowable limits but these were assessed and shown to be acceptable. Monitoring of duct displacements and strains and hanger loads revealed no unexpected behaviour.

7. PLANT REINSTATEMENT

Following repair and PWHT, the pressure circuit and boiler internals were reinstated to perform their normal operational functions.

The procedure for reinstating the top duct bolted flange connection above the main gas valve was essentially the reverse of that for separation. The top duct nozzle forces and moments were restored to their pre-separation values ensuring that the ducts and nozzles were suitable for return to normal service.

The superheater tube banks were reinstated on their support brackets to ensure that they and their support arrangement would be suitable for long term operational service. Following reinstatement a hydraulic overpressure test was carried out on the tube banks to provide additional confidence in their continued operational integrity.

8. PLANT MONITORING DURING REPAIR ACTIVITIES

Extensive monitoring undertaken during the site activities included strict control of all welding parameters during the repairs as well as displacement and strain monitoring of the ducts, boiler inlet nozzles and hangers during separation and PWHT. Widespread temperature monitoring of the boiler shell, attachments and internal components and displacement monitoring of tube banks was also undertaken during PWHT. This monitoring provided assurance that all repair welding was satisfactory, that the specified PWHT heating rates and temperatures had been achieved, that the superheater tube bank support brackets remained unloaded and that no unexpected duct behaviour had occurred.

9. INSPECTION

Extensive and rigorous inspections of the repaired and profiled welds were undertaken using both magnetic particle and ultrasonic techniques to provide assurance that the required quality to sustain all normal operational and postulated fault loadings had been achieved.

The principal objectives of these inspections was to:-

1. confirm the satisfactory condition of the ligaments following defect removal and weld preparation of the Course 6/7 repair sites and following profiling of the Course 5/6 defect,

2. confirm that the Course 6/7 repair sites were sound after welding and

3. confirm that no significant defects were introduced to either the Course 5/6 or Course 6/7 sites during PWHT.

The inspections met or exceeded the requirements of ASME (1). A combination of magnetic particle inspection, manual and automated ultrasonic inspection provided reliable, diverse and redundant inspection. The inspections were qualified appropriately following the general methodology of the European Network for Inspection Qualification (ENIQ). The qualification was based on technical justification and included test block trials.

The inspection results following weld preparation of the Course 6/7 repair sites demonstrated that the sites were in a satisfactory condition for welding. Inspections undertaken following welding and prior to PWHT revealed only a few minor welding defects and these were all of no more than 3mm in through wall extent, the approximate limit of NDT detection capabilities. No local repairs were required on any boiler. Further inspections were undertaken following PWHT and these identified no evidence of defect initiation or growth during PWHT.

Inspection of the Course 5/6 weld following final profiling demonstrated that all significant defect indications had been removed. No defect indications with a through wall extent in excess of 3mm remained in the profiled weld.

For all other welds and components which were subjected to high temperatures during PWHT, i.e. those within about 2m of the repair weld, such as shell axial welds, thermal sleeve attachment welds and boiler beam support bracket welds, future integrity was underwritten by extensive inspections. These inspections, principally to confirm the absence of any significant degradation during PWHT, were carried out before and after PWHT. They were undertaken using approved Magnox Procedures and were subject to Capability Reviews. The results of the pre-PWHT inspections demonstrated that the features were in a suitable condition for PWHT to commence and the post PWHT results showed no evidence of defect initiation or growth during PWHT.

10.   REFERENCE DEFECT ASSESSMENTS

Although the integrity of the repaired and profiled welds was generally underwritten by code compliance, quality control and inspections, support was also provided by the results of fracture mechanics assessments carried out using the R6 methodology (2). Satisfactory reserve factors were demonstrated, using lower bound material properties, for extended reference defects of 6 mm through wall extent. The reference defect size was selected on the basis that it provided a factor of at least 2 on the NDT detection capabilities. However sensitivity studies demonstrated that the results were not particularly sensitive to defect size and substantially larger critical defect sizes were demonstrated.

11.   FOREWARNING OF FAILURE

A dedicated and sensitive CO_2 detection system was installed on each of the repaired and profiled welds prior to return to service. It is capable of detecting leakage through fully penetrating defects of much less than the critical length. This system will remain in service at least until the first statutory outage of the reactor when a full reinspection of the repaired and profiled welds will be undertaken.

The reactor was shut down and reinspections were undertaken on the repaired and profiled welds after a short period of operation. The inspections were undertaken using both ultrasonic and magnetic particle techniques. The results obtained confirmed the high quality of the repaired and profiled welds and showed no evidence of any in-service defect initiation or growth.

12.   CONCLUSIONS

The integrity of the repaired and profiled welds was underwritten by the application of rigorously controlled and code compliant repair techniques, extensive inspections and design/fracture assessments.

It was demonstrated that the integrity of other pressure circuit components was not significantly affected by the repair and PWHT activities by a combination of careful control and monitoring and extensive NDT inspections.

13.   ACKNOWLEDGEMENT

This paper is published with the permission of the Director Technology and Central Engineering, BNFL Magnox Generation..

14.   REFERENCES

1   American Society of Mechanical Engineers Boiler and Pressure Vessel Code, ASME XI, 1989.

2   I Milne, R A Ainsworth, A R Dowling and A T Stewart, Assessment of the Integrity of Structures Containing Defects, CEGB Report No R/H/R6 Rev 3, including updates to February 1997.

3   BS5500:1997, Specification for Unfired Fusion Welded Pressure Vessels.

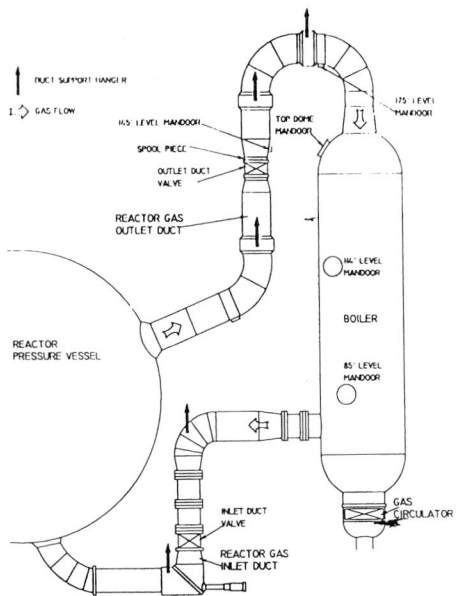

Figure 1: Schematic layout of gas circuits at Sizewell A

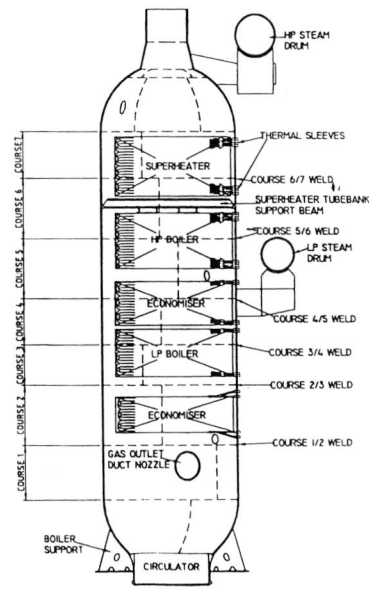

Figure 2: Schematic layout of boiler at Sizewell A

S690/007/99

Inspection challenges

K J BOWKER
BNFL Magnox Generation, Berkeley, UK

ABSTRACT

The weld repairs of the Sizewell 'A' boilers required an extensive programme of non-destructive testing (NDT), which encompassed both the repaired welds and a range of other welds, including those that could be adversely affected by the post weld heat treatment. Great emphasis was placed on the NDT because there was no pressure test on completion of the repair. Therefore, the capability of the NDT had to be shown to be of the highest standard. To this end, the inspections were formally qualified by an independent body and, to provide further confidence in the reliability of the inspections, both manual and automated inspections were applied to the repaired welds. Thus, there was an extensive development programme and one inspection featured prominently. This was an inspection to monitor for possible stress relief cracking resulting from post weld heat treatment for which the data had to be collected at $\sim 160\,^{\circ}$C.

This paper addresses the scope of the inspections, the development work and site implementation.

1. INTRODUCTION

All Magnox Nuclear Power Stations have an extensive and ongoing inspection programme. During 1996, inspection of the Sizewell 'A' boiler shell welds revealed significant defects in four circumferential welds(1). BNFL Magnox took the decision to repair three of the welds by excavating the defects and rewelding followed by stress relief. The fourth was repaired by removing the defect and profiling the repair so that stresses were reduced to an acceptable level.

The boiler shell weld repairs included an extensive programme of Non-Destructive Testing (NDT) before welding, after welding and before stress relief and finally after stress relief to ensure that there were no defects of safety significance and that the boiler could safely be returned to service. Furthermore, the Nuclear Installations Inspectorate (NII) was concerned about the possibility of stress relief cracking during the heat treatment. Consequently, it was necessary to show with a high degree of confidence that there had been no change or defect growth during the heat treatment. The fingerprint inspection before heat treatment had to be carried out at high temperature, ~150-170 $^\circ$C.

In the absence of a proof pressure test after the repair, the NDT had to be shown to be of a very high standard. Consequently, it was agreed with the NII, that several inspections of the repair welds would be qualified following the European Network for Inspection Qualification (ENIQ) methodology(2).

This paper addresses the scope of the inspections, the development programme with the challenges that had to be overcome and the implementation of the inspections on site.

2.   INSPECTION SCOPE AND OBJECTIVES

The primary objective of the inspections was to provide an extremely high level of confidence that, following the repair, the boiler was clear of significant defects and could be safely returned to service without a proof pressure test. As shown schematically in Figure 1, the boiler is constructed from cylindrical courses, 57 mm thick, ~7 m external diameter, between 2.1 and 3.2 m high, and made from three 120^0 sections welded together axially. Not shown on the figure are the various attachments welded to the boiler shell, such as the boiler beam support pads. Boiler tubes penetrate the boiler shell through thermal sleeves which are welded to the boiler shell.

For three boilers, the full circumference of the course 6/7 welds were repaired by welding and ~1.6 m of the course 5/6 weld in boiler 2C were repaired by removing the defect and profiling to reduce stresses. The definitive inspections of these repairs were carried out on completion of the repair, after stress relief by post-weld heat treatment (PWHT), to ensure that the repaired welds were of good quality, that there were no safety significant defects present and that there had been no significant stress relief cracking resulting from the PWHT. However, during the course of the repair there were inspections at various stages. The machined course 6/7 repair weld prep and the remaining ligament were inspected to ensure that the major defects had been completely removed and that there were no other defects which would adversely affect the success of the weld repair. After welding and before PWHT, the repair weld was inspected to ensure that there were no defects requiring repair before the PWHT. In addition, there was the requirement to demonstrate that there had been no significant stress relief cracking resulting from the PWHT, and therefore fingerprint inspections were carried out after welding but before PWHT and again after PWHT, which were compared to show that there had been no significant change.

The stress relief of the course 6/7 repair welds increased the temperature of the boiler shell within ~2 m of the repair welds above the normal operating temperature and introduced thermal stresses. Consequently attachments welds in this region were inspected before PWHT to ensure that they were of sufficient quality and after PWHT to ensure that no safety significant defects had resulted from the PWHT and that there had been no major degradation. The axial seam welds in course 6 and 7 were also inspected but, because both courses 6 and 7 are ~2.4 m high, there was no requirement to inspect the adjacent circumferential welds.

During the course of the repair programme, many pipes and tubes were cut, which subsequently had to be rewelded. These welds were inspected using standard techniques to ensure that they were fit-for-purpose.

Thus, inspections were carried out on a range of welds at particular stages during the repair. Table 1 shows a limited selection of the welds inspected at various stages of the repair. The table does not include the initial inspections which found, characterised and sized the defects, nor the in-service inspections which will be carried out at the first outage following the repair. The purpose of the inspections at each stage of the repair is summarised. Before PWHT, the purpose is essentially commercial, to help ensure the success of the repair. For example, the manual ultrasonic inspection of the course 6/7 weld after welding but before PWHT searched for any structurally significant defects or any which might grow during PWHT. If the inspection had found any such defects they could have been repaired before PWHT. Inspections after PWHT are safety related, ensuring that the boilers can safely be returned to service.

The repair was implemented in accordance with the ASME pressure vessel code (3, 4, 5) with justification of any non-compliances. For the inspections, compliance with ASME in terms of scope, acceptance standards, procedures, personnel and performance demonstration (qualification) was documented with justification of the few non-compliances. A substantive non-compliance was the ASME requirement for examination of the initial layer of the repair weld metal by MPI. This was not carried out because of the danger of contamination from the MPI adversely affecting the weld. Since there was a full volumetric inspection of the completed weld, it was accepted that there were no safety implications in omitting this inspection.

The reporting criteria for the inspections needed to be consistent with the ASME acceptance criteria, which they were. Furthermore, for the final inspections of the repaired boiler seam welds after PWHT there was a substantial margin between the smallest defect size set for reliable detection and the critical defect size. The smallest defect size for reliable detection was 3 mm through wall by 20 mm long or 5 mm through wall by 12 mm long. This compares with a reference defect size of 16 mm through-wall by 250 mm long. For the Sizewell 'B' pressurised water reactor (PWR) pressure vessel the concept of a validation factor was introduced, which was the ratio of the reference defect dimensions to those of the minimum defect size for reliable detection. This was required to be 2 (i.e. 4 times the area). The validation factor for the Sizewell 'A' boilers is »2 and considerably higher than that required for the PWR pressure vessel. This provides great confidence that any safety significant defect would have been found and reported.

3.   INSPECTION STRATEGY AND PROGRAMME

As discussed above, inspections were carried out at various stages of the repair programme and were not confined to the repair welds. The list of welds inspected is much more extensive than that shown in table 1 and includes attachment welds other than thermal sleeves affected by PWHT, boiler tubes and welds and the inlet cone weld for boiler 2C. The table includes the inspection techniques that were applied. For most inspections of the course 6/7 repairs, there was surface inspection using magnetic particle inspection (MPI) and volumetric inspection with both manual ultrasonics (MUT) and automated ultrasonics (AUT). Both manual and automated ultrasonic inspections were applied to provide added confidence in the inspections because of their importance. Applying two independent inspections adds diversity and redundancy which increases the confidence that no significant defects will be missed. Manual inspection is flexible and relatively quick, whilst automated inspection provides permanent records which can be independently reviewed and which can be compared with subsequent inspections.

To increase confidence in the most important of these inspections further, it was agreed with the NII that they should be formally qualified. Inspection qualification is a systematic assessment of an NDT system to ensure that it is capable of achieving the required performance under real inspection conditions. It was agreed with the NII that the qualification would be carried out following the European methodology, ENIQ (2). This methodology which covers procedures, equipment and personnel, is implemented by a qualification body and utilises a technical justification and practical trials as appropriate. The qualification body comprised BNFL Magnox staff and external contractors and was overseen by an independent qualification body (IQB) from the AEA-T Inspection Validation Centre (IVC) who had been responsible for qualification of inspection of incredibility of failure components for the Sizewell 'B' pressure vessel. On satisfactory completion of a qualification, certificates were issued by the IVC.

Two levels of qualification were applied to these inspections and these are shown in table 1. Those denoted IQB were qualified using technical justification and open trials, which use a representative test specimen containing known defects to confirm the conclusions of the technical justification and demonstrate that the procedure and equipment work in practice. The automated inspections monitoring for growth (AMG) were sufficiently novel that it was agreed that the data analysts should be individually tested to show that they could evaluate growth satisfactorily. To this end the operators were required to analyse data without knowledge of the defects. These blind trials were in addition to the technical justification and open trials required for the procedures and equipment, and this qualification is denoted by IQB BT in Table 1.

Increased confidence was also required in the inspection of certain attachments, including the thermal sleeves and boiler beam support pads. For these, capability statements were prepared which were reviewed externally by Rolls-Royce Marine Power (RRMP) who endorsed the statements when they were satisfied.

Thus, as part of the development programme, it was necessary to prepare technical justifications or capability statements for the inspections and, where appropriate, to demonstrate the performance to the IQB so that a certificate could be issued, before commencement of the inspection on site. Qualification of these inspections is reported fully by Baborovsky (6).

4.   INSPECTION DEVELOPMENT

As stated in section 2 above, the primary purpose of the inspections was to demonstrate with an extremely high level of confidence that, following the repair, no significant defects were present. Whilst, there was generally a substantial margin between critical defect size and the generally accepted capability of the inspections, procedures had to be developed which would ensure the inspections were carried out to a high standard and were repeatable. Repeatability is an issue when the results of successive inspections would be compared, and this was especially important when monitoring for possible defect growth resulting from PWHT. Other major issues for NDT procedures were problems arising from geometry and access, high temperatures and the novel application of techniques.

Whilst some development of techniques was required for MPI, these were relatively straightforward and are not discussed here. The design of ultrasonic techniques was more challenging. For conventional pulse-echo testing, the key issue is to ensure that the full inspection volume is scanned by beams of appropriate angles and with sufficient sensitivity to ensure that all plausible defects above the required size will be detected. As illustrated in Figure 2, a beam of ultrasound is transmitted into the steel at a particular angle and is scattered back from any discontinuity. Clearly, planar defects will be most reliably detected if the ultrasonic beam is approximately normal to the plane of the defect. However, at oblique incidence, ultrasound is scattered back from the edge of the defect, and if the face is rough, from the face of the defect. With the probes and sensitivities specified in the BNFL Magnox Generation company standard procedures, defects can generally be detected at angles up to $\sim 30^0$ away from normal incidence. With the exception of the automated monitoring for growth (AMG), all ultrasonic inspections, including the final pulse echo inspections after PWHT, were based around the company standard procedure. The development of the AMG inspections was by far the most challenging. The development of this inspection, and three others are described briefly below. In general, problems to be overcome fell into four main categories, geometry and access, repeatability, high temperature and novel application of techniques.

4.1 Thermal Sleeves Attachment Welds

Boiler tubes pass through the thermal sleeves where they penetrate the boiler shell. They are in banks of four closely spaced rows as illustrated in Figure 1, and penetrate the boiler shell at angles between 0^0 and 42^0. A thermal sleeve and attachment weld is shown schematically in Figure 3.

Thermal sleeves close to the repaired welds would be subject to high temperatures and thermal stresses during PWHT and failure at the attachment welds was seen as a threat to the repair. Consequently, before commencing the repair, it was decided to inspect the bank of thermal sleeve attachment welds closest to the course 6/7 seam to discover whether the attachment welds were of sufficient quality that the PWHT would not introduce unacceptable defects. They were also inspected after the repair to ensure that after PWHT, there were still no unacceptable defects.

Geometry and access made inspection of the thermal sleeves difficult. The BNFL Magnox Generation company standard MPI procedure could, however, with minor adaptation, be used for surface inspection of the weld and the ligament. However, it was necessary to develop a volumetric inspection to detect both circumferential defects such as lack of sidewall fusion or HAZ stress relief cracking in the boiler shell and transverse defects in the radial axial plane of each thermal sleeve.

For volumetric inspection, radiographic inspection was eliminated because it would not detect many of the defects of concern, which left ultrasonic inspection. There is insufficient space between adjacent thermal sleeves for meaningful scanning from the surface of the boiler shell. Scanning from the outer surface thermal sleeves was not practicable because it would be necessary to rotate the probe in a complex fashion as it was scanned round and along the sleeve in order to inspect the weld at appropriate angles, e.g. to detect a lack of side-wall fusion defect. In addition, for some angles of thermal sleeve, significant volumes of the weld could not be inspected. Finally, for the attachment weld on the inside surface of the boiler shell, the stub of the thermal sleeve was not long enough to enable the weld to be inspected. Consideration was given to inspecting from the bore of the thermal sleeve by inserting special ultra-thin probes down the annulus between the thermal sleeve and the boiler tube. This would not have had the required capability to detect the full range of defects of concern and, to insert the probes, it would have required all the welds to the tubes, internal to the boiler to be removed, and subsequently re-made.

The final option was to inspect the welds from the weld cap using subminiature probes. It was shown that, scanning with 45^0, 60^0 and 70^0 probes pointing in both circumferential and both radial directions, the inspection would have the required capability, except for a few areas on some thermal sleeves. However, this approach required the weld surface, which was in the as welded condition, to be dressed smooth. A number of thermal sleeves representing a range of access restrictions and sleeve angles were selected to assess feasibility of hand dressing and inspecting. These trials were successful and consequently the inspection was implemented. All the external welds were dressed and inspected, but due to access problems, not all the internal welds could be dressed and inspected.

Suitably qualified NDT operators were trained and required to demonstrate their competence on a test specimen containing defects, before inspecting the thermal sleeve welds. Capability statements were prepared for both the ultrasonic inspections and the magnetic particle inspections, which set out the detection and sizing capability and limitations. These statements were reviewed and endorsed by RRMP, which gave confidence that the inspections would have detected any significant defects of concern.

No defects of concern were found before the repair which provided confidence that failure of the thermal sleeves during PWHT was unlikely to be a problem. This was confirmed when no defects of concern were found following PWHT.

4.2 Course 6/7 Weld After PWHT

This was the definitive inspection to show that the boilers were safe to return to service. As such, the capability had to be demonstrated to the highest level of confidence. However, in terms of ultrasonic techniques it was a very straightforward inspection, being essentially a standard inspection of a double 'V' butt weld. Both manual and automated inspections were carried out. The automated inspections, like all the automated inspections for the repairs, used the in-house MIPS and GUIDE data collection and display software. MIPS controlled data collection using the Rolls-Royce Deltapulse ultrasonic system.

The weld preparation for the repair is shown in Figure 4. As can be seen, the repair weld is wider than a standard butt weld and there is a flat base. The width of the weld repair increased locally to a maximum width of ~110 mm. The transition to the increased width was with a taper of ~11 .

The plausible planar defects of concern included, lack of sidewall fusion, stress relief cracking in the heat affected zones of both the repair and the original weld, centre-line cracking and hydrogen cracking normal to the boiler shell surface in both the repair weld and the original weld. The taper, described above, at the change in width of the excavation would affect the orientation of defects associated with the weld prep, such as lack of sidewall fusion and cracking in the HAZ. Such defects would be skewed at ~11^0 relative to the weld centerline. Normally probes are scanned with their beam axes perpendicular to or parallel to the weld centerline to detect longitudinal and transverse defects respectively. However, in order to maximise the capability to detect defects in the region of the tapers, the probes were scanned both perpendicular to the weld centerline and skewed appropriately so as to provide near normal incidence for defects parallel to the tapers.

Test blocks were manufactured containing representative plausible defects and trials conducted to provide evidence for the technical justification. The technical justification defined the 'worst case defects' for detection and mathematical modelling demonstrated that these could be detected. The technical justification took into account factors such as residual stresses which can affect defect detectability and demonstrated reliable detection of all plausible defects of concern. The technical justification was accepted and approved as part of the inspection qualification.

Open trials of both the manual and automated procedures were carried out with test blocks containing representative defects including worst case defects. On successful completion of the trials, a qualification certificate was issued, and a copy of the front side of that for the manual inspection of this weld is reproduced in Figure 5. The certificate for the automated inspection is similar. Before the inspections could be undertaken at site, all personnel had to be approved by the qualification body as having the appropriate qualifications and familiarisation with the procedures and equipment. The inspections were then carried out at Sizewell.

4.3 Boiler 2C Course 5/6 Weld Profile After PWHT

The defect in the boiler 2C course 5/6 weld was shallower than those in the three course 6/7 welds that were repaired. This weld was repaired by machining out the defect with a profile that kept stresses down to an acceptable level. This profile had to enable a thorough, high quality inspection of the remaining ligament by NDT. The final profile that was arrived at had a flat base and sides that sloped at 30^0 as shown in Figure 6. This profile was wide enough to enable much of the weld to be inspected from the base of the excavation. The remainder was inspected from the original surface of the boiler shell outside the excavation. The 30^0 slope for the sides meant that the sides did not cause any significant gaps in the coverage for defects.

One problem particular to this inspection, was that there was known to be a mismatch between the surfaces of courses 5 and 6. Whilst the weld had been dressed, it was felt that detection using beams reflected from the inside surface of the boiler shell could not be relied on because error of form of the surface could misdirect the beam. This was only a problem for defects normal to and very near to the surface of the base of the excavation, but not surface breaking. (Surface breaking defects would have been detected by MPI.) Using ultrasonics, such defects are generally detected because they give a very strong signal by the "corner effect" with a 45^0 beam reflected off the back surface. Effectively, the defect and the adjacent surface form two mutually orthogonal mirrors and the sound beam is reflected back along its original path after striking both mirrors.

Due to the mismatch in surfaces, the corner effect mechanism could not be relied on in this case, and it had to be shown that such defects could be reliably detected directly from the base of the excavation rather than relying on sound reflected from the inside surface. Generally, unless the surface finish is of very high quality and the probes are of high quality, surface noise can obscure signals from such defects. However, experimental trials demonstrated that for this inspection the probes and surface were of sufficiently high quality that such defects could be reliably detected directly. The 45^0 probe beam reflecting sound from the inside surface was still used since it would probably detect the defects, but the technical justification and the qualification did not require the use of this technique.

A further problem with the course 5/6 was how to end the excavation since, unlike the course 6/7 repairs which were fully circumferential, this repair was of limited length. The initial plan had been to round the ends off with semi-circular sidewalls as shown in Figure 7a. However, this would have made the ends of the excavation very difficult to inspect and it would have been impossible to reliably inspect the area under the semicircular sidewalls with their 30^0 slope. Consequently, a flat run-out was selected as shown in Figures 7(b) & 7(c)c. This resulted in a longer excavation but did not present any inspection problems.

The inspection procedure was qualified in the same way as that described for the inspection of the course 6/7 welds after PWHT. The operators were also approved by the qualification body in the same way.

4.4 Course 6/7 Weld - Monitoring Defect Growth

There was concern about the possibility of stress relief cracking resulting from PWHT and an inspection was required that would identify stress relief cracking. It is not possible to distinguish stress relief cracking from other types of cracking based on its ultrasonic response alone. However, by comparing the results of inspections before and after PWHT and monitoring for change, differences outside those attributable to measurement error could be attributed to changes due to PWHT. These changes would therefore probably be stress relief cracking. The objective, therefore, was to design a highly repeatable inspection that would reliably detect change. This inspection became known as AMG - automated monitoring for growth.

The initial concept was to compare time-of-flight diffraction (TOFD) data from before and after PWHT. TOFD is illustrated in Figure 2. A pair of probes with relatively wide beams is scanned over a defect and sound is diffracted from the defect edges to the receive probe. The travel time (time-of-flight) of the diffracted signal gives the distance travelled and knowing the probe separation, the depth of the defect edge below the scanning surface can be readily calculated. If the probes are re-scanned at the same locations, then successive TOFD inspections provide very repeatable data.

TOFD data can be difficult to interpret without some knowledge of the defects that are present. However, the pulse-echo inspections of the welds after PWHT would provide the necessary data on any defects to enable the TOFD data to be interpreted reliably. The first set of AMG data must be collected before PWHT when defect locations are not known. It is therefore necessary to collect fingerprint data from the full extent of the repair, i.e. the full weld circumference. This also requires pairs of probes with different separations to cover defects at different depths through the boiler shell, and these pairs need to be scanned at a range of locations across the weld and HAZ to ensure coverage of the full inspection volume.

The requirements for permanent records and highly repeatable scanning led to an automated inspection. Repeatability was enhanced by mounting the scanner on studs permanently welded to be boiler shell. These studs were also used as datums for all the inspections of the circumferential welds.

4.4.1 Inspection at elevated temperatures

Metallurgical and materials property considerations meant that, after completion of the welding, the temperature could not be allowed to drop significantly below the pre-heat temperature before PWHT. Consequently, the fingerprint before PWHT had to be carried out at a temperature of ~150°C or higher. Ultrasonic inspection is generally done at ambient temperature, and the problems start to arise at temperatures above ~50°C. Note that there were two separate ultrasonic inspections carried out at ~160°C, the AMG inspection described here and the manual ultrasonic inspection of the course 6/7 weld carried out after welding and before PWHT. The inspection procedure for the latter was developed by Mitsui Babcock

Technology Centre who had gained experience of high temperature ultrasonics through a European Commission development project concerned with NDT during welding (7).

There are four principal issues for high temperature ultrasonic inspections which are described briefly below.

- Conventional transducers are designed to operate at ambient temperatures. At temperatures of only ~50°C the probe shoe material can start to deteriorate. At higher temperatures, the sensitivity of the piezo-electric transducer elements starts to fall as the Curie temperature is approached. There can also be problems with the adhesives used to attach the transducer to the probe shoe.

- The velocity of ultrasound in the probe shoe and in the steel changes with temperature. The change of velocity in steel is small and does not significantly affect the calculated distances travelled by the ultrasound computed from the travel times. However, angle beams are generated by diffracting sound from the probe shoe into the steel. Consequently, the change in velocity in the probe shoe material can result in significant changes in the beam angle.

- As the temperature rises there are losses in sensitivity and signal-to-noise ratio. These can be ameliorated by using lower frequency ultrasound but this reduces resolution and hence accuracy.

- At high temperatures, the choice of the liquid couplant that enables the ultrasound to be transmitted from the probe into the steel is problematic. If the liquid approaches its boiling point then bubbles can form which scatter the ultrasound, reducing sensitivity and increasing noise. Thus aqueous based couplants are not suitable once the temperature approaches 100 ° C. Other liquids, including those with an organic base may start to decompose and give off harmful vapours.

Special high temperature probes were commissioned from two manufacturers, and following trials one manufacturer was selected to provide the probes for the inspections. These had a special probe shoe material that did not soften and in which the velocity of sound was less dependent on temperature than in conventional materials. Despite this, probe beam angles still changed by ~5 as the temperature was raised from ambient to 150 ° C. The lead metaniobate transducer elements were mechanically clamped to probe shoes. The probe frequency was reduced from 5 MHZ to 3.5 MHZ. The probes were designed to operate satisfactorily at temperatures up to 170°C

A number of possible liquid couplants were investigated. Since it would not be possible to remove all traces of couplant after the inspection, the couplant had to withstand the PWHT without decomposing into noxious chemicals. A silicon oil based couplant was selected. This had to undergo a COSHH assessment before use.

Finally, the scanner had to be designed to operate with high precision at temperatures in excess of 150°C without any components degrading or being affected by differential expansion. Care had to be taken with motors and encoders and ultrasonic signal pre-amplifiers had to be air cooled.

4.4.2 Inspection Procedures

Development trials showed that the TOFD data was highly repeatable, even scanning at the elevated temperature. However, for practicable configurations, TOFD did not have the required capability close to the scanning surface. In particular, the lateral wave which travels directly along the surface between the two probes, obscured signals from near surface defects and so their edges could not be reliably detected. This problem was exacerbated by the reduction in frequently. Consequently, pulse-echo inspection had to be used for the near surface region.

Unlike TOFD, which uses travel time alone to compute the depth of an indication, pulse echo uses both the range and the beam angle to calculate the location. Consequently any error in the beam angle, such as that caused by a change in temperature, will result in an error in location. However, the error increases with range, i.e. with defect depth, and will be small for near surface defects. Development trials showed that satisfactory repeatability and accuracy could be obtained using pulse-echo for near surface defects.

In principle, for TOFD alone, it might have been possible to achieve sufficient repeatability if the inspections before and after PWHT had been at different temperatures. However, the significant change in performance of the probes between ambient and 150°C made doing both inspections at the same temperature highly desirable. The introduction of the pulse-echo tests meant it was essential that they should be at the same temperature in order to minimise the effects of change in beam angle on the pulse-echo results. Consequently, after PWHT, the boiler was maintained at ~150°C whilst the second set of AMG fingerprint data was collected, before it was cooled to ambient. Another AMG fingerprint was collected at ambient temperature for comparison with future in-service inspections.

As the inspection was being formally qualified, it was necessary to define a rigorous set of rules for the data analyst to determine whether in any given situation pulse-echo or TOFD data should be used. Without this, the procedure could be applied differently by different analysts and hence it would not be possible to ensure that the procedure would always work correctly and hence it could not be qualified. To this end, it was necessary to define three depth zones. There was a near surface zone where pulse-echo was always used, a deep zone where TOFD was always used and an intermediate zone where the technique used was defined by a set of rules whose objective was to ensure that wherever possible, the same technique was used for any given defect edge before and after PWHT, and if possible the same technique was used for both defect edges before and after PWHT.

In addition to procedure qualification, as this was a novel application and the data analysis was unusual and complex, the data analysts were individually qualified. The analysts underwent specific training for TOFD data analysis and for the AMG data analysis procedure. They had to pass a written examination on the procedure and the rules for the technique to be applied and, in blind trials, they had to analyse data from test blocks to assess defect growth. Five out of nine candidates passed and were qualified.

5.   SITE IMPLEMENTATION

The inspections whose development is described in section   above were some of the key inspections which required significant resource for their development, site implementation and, where appropriate, qualification. However, they constituted only a small fraction of the total number of inspections carried out during the repair programme. A large number of inspections, which had not initially been anticipated, arose during the course of the repair. For example, there were inspections of temporary structural steel work which supported the boiler tubes during PWHT. An NDT work control system was introduced to co-ordinate all the inspections so that the necessary resource would be available when required.

The scale of the challenge of successfully co-ordinating the NDT programme can be seen since there were more than 1400 inspection reports associated with the repair programme and for the automated inspections, the data was archived onto over 360 CD ROMS. This is in addition to the 36 CD ROMS of archive data from the development work and qualification trials. Every inspection was on a quality plan and the major inspections were each the subject of their own quality plan. Manual inspections were subject to audit by re-inspection or surveillance where re-inspection was impracticable whilst, for the automated inspections, the data was analysed twice, independently by different analysts. Every inspection report was verified and the results were entered in a database for review by the project defect assessment panel.

All the inspection personnel were appropriately experienced and qualified. Thus, for example, all manual ultrasonic personnel were certified to PCN level 2, and personnel who may have been required to measure defect size had the PCN critical sizing module. Inspection personnel for the automated inspection had been trained and certified by the in-house MIPS and GUIDE schemes for data collection and analysis respectively. Where appropriate, personnel were provided with additional training. For example, operators were given specific training for the hot manual inspections of the course 6/7 welds after welding and before heat treatment. Specific safety requirements were included in the procedures for this inspection and for the hot AMG. Also, for these hot inspections, special precautions were taken to minimise the risk of couplant dripping onto the heat pads. Drip trays were tack welded to the boiler and sealed with a high temperature sealant and protective blankets covered the heater pads.

The inspections were all developed and, where necessary, qualified by the time they were required for the programme. For the automated inspections, durations decreased with successive applications of the procedures as any site teething problems were overcome. Analysis times improved significantly when a direct network link was set up between the data acquisition equipment inside containment and the data analysis portacabin with the computer workstations.

In order to minimise inspection time and keep to programme, there was 24 hours/day manning. This required a large team of NDT personnel. For the development and site application of the automated inspections, the project used most of the available resource of personnel with the necessary training, experience and certification.

With such a large inspection programme, it was important to ensure that every inspections was implemented and reported correctly. There was a full audit programme for both implementation and reporting. In addition, Independent Third Party Inspection Agency checked rigorously that the inspections were carried out and reported correctly in accordance with the quality plan and the inspection procedure, with calibrated equipment and by suitably qualified personnel.

The inspections were carried out successfully and demonstrated that the repair had been carried out satisfactorily and the boilers were returned to service. After operation for about one month, the boilers were removed from service and the repaired welds on all three boilers were inspected manually at ~90 ° C in a single day. There was no evidence of degradation and the boilers were again returned to service.

6. CONCLUSIONS

- Great emphasis was placed on the NDT because of the absence of a proof pressure test and hence the NDT had to be of very high quality.

- High confidence in the NDT capability was provided, where appropriate, by inspection qualification and by multiple inspections such as manual and automated NDT.

- The programme for inspection development and qualification was challenging.

- All the challenges were overcome and the inspections were developed and where appropriate qualified for the required programme.

- NDT was a very major activity which required significant resources.

- The inspections were completed to the required programme.

- The inspections met the requirements set to demonstrate that the boilers complied with the safety arguements.

7. ACKNOWLEDGEMENT

The development, qualification and site implementation of the inspections was a team effort. The team was very large and it is not possible to acknowledge everyone who played a significant role, however particular thanks are due to:

- Colin Bird, Mark Kubulus, David Wood and Martin Yates of BNFL Magnox

- Ian Atkinson, Terry Bann, Brian Hawker and Nick Turner of AEAT

- Fraser Hardie, Ian McCormick and Mike Venters, of Mitsui Babcock

- Alistair Pink and Peter Kelsey of Rolls-Royce Marine Power

- David Denby and Chris Green of QVS

- Colin Bristow, Hughie Elder, Pete Tuck and George Yates of SENDT

- Geoff Ellis, Derek Skelton and Trevor Stamford of OIS

- All the other staff from CET Medway, OIS, Mitsui Babcock, SENDT, Rolls-Royce MP and ABB who were members of the site inspection teams.

This paper is published with the permission of the Director Technology and Central Engineering, BNFL Magnox Generation.

8. REFERENCES

1   Exworthy, LF and Flewitt, PJ, Sizewell 'A' boiler repair: detection and diagnosis of defects, paper 2 of these proceedings

2   European methodology for qualification of non-destructive testing (second issue), ENIQ report No 2, EUR 17299 EN, 1997.

3   ASME Boiler and Pressure Vessel Code; Section XI, Rules for Inservice Inspection of Nuclear Power Plant Components, 1995 Edition including 1996 addenda.

4   ASME Boiler and Pressure Vessel Code; Section V, Nondestructive Examination, 1995 Edition including 1996 addenda.

5   ASME Boiler and Pressure Vessel Code; Section III, Rules for Construction of Nuclear Power Plant Components, Division 1 - Subsection NB Class 1 Components, 1995 Edition including 1996 addenda.

6   Baborovsky, VM, Sizewell 'A' boiler repair: NDT qualification, paper 11 of these proceedings.

7   EC Contract No. BRE 2-CT92-0319, NDT methods for flaw detection during welding.

TABLE 1 - SIZEWELL A BOILER REPAIR INSPECTIONS (examples)

Stage	Purpose	Course 6/7 Repairs		Course 5/6 profile (2C only)		Thermal sleeve attachmnt welds and ligaments	
		Tech-niques[1]	Quali-ficatn[2]	Tech-niques[1]	Quali-ficatn[2]	Tech-niques[1]	Quali-ficatn[2]
Before starting repair	Commercial: Integrity prior to starting repair: no major defects			None	n/a	MPI	
		MUT				MUT	
		AUT					
During Excavation	Commercial: Confirm defects removed	MPI		MPI			
After Excavation to weld profile (6/7 welds only)	Commercial: Integrity of remaining material before welding	MUT	IQB				
		AUT	IQB				
After Excavation to intermediate profile (2C 5/6)	Commercial: Integrity of remaining material			MPI			
				MUT			
After welding, before PWHT	Commercial: No major defects requiring repair	MPI					
		MUT					
		AMG	IQB BT				
After PWHT	Safety: Fitness-for-purpose. *For repaired welds: No significant degradation from PWHT*	MPI		MPI		MPI	CS
		MUT	IQB	MUT	IQB	MUT	CS
		AUT	IQB	AUT	IQB		
		AMG[3]	IQB BT				
After Pressure cycle	Safety: Fitness-for-purpose. *No significant degradation.*	MPI		MPI		None	n/a
		MUT		MUT			

Notes: [1] MPI:- Magnetic Particle Inspection,
MUT:- Manual Ultrasonic Testing,
AUT:- Automated Ultrasonic Testing
AMG:- Defect Growth Monitoring - Time of Flight Diffraction plus pulse-echo for near surface zone

[2] IQB:- qualified with independent qualification body,
IQB BT:- qualified with independent qualification body including blind trials for personnel,
CS:- capability statement endorsed by independent third party

[3]Defect growth monitoring repeat inspections are carried out at the same temperature as the initial inspection. There are be two scans, one hot for comparison with before PWHT, one cold for comparison with subsequent in-service inspection.

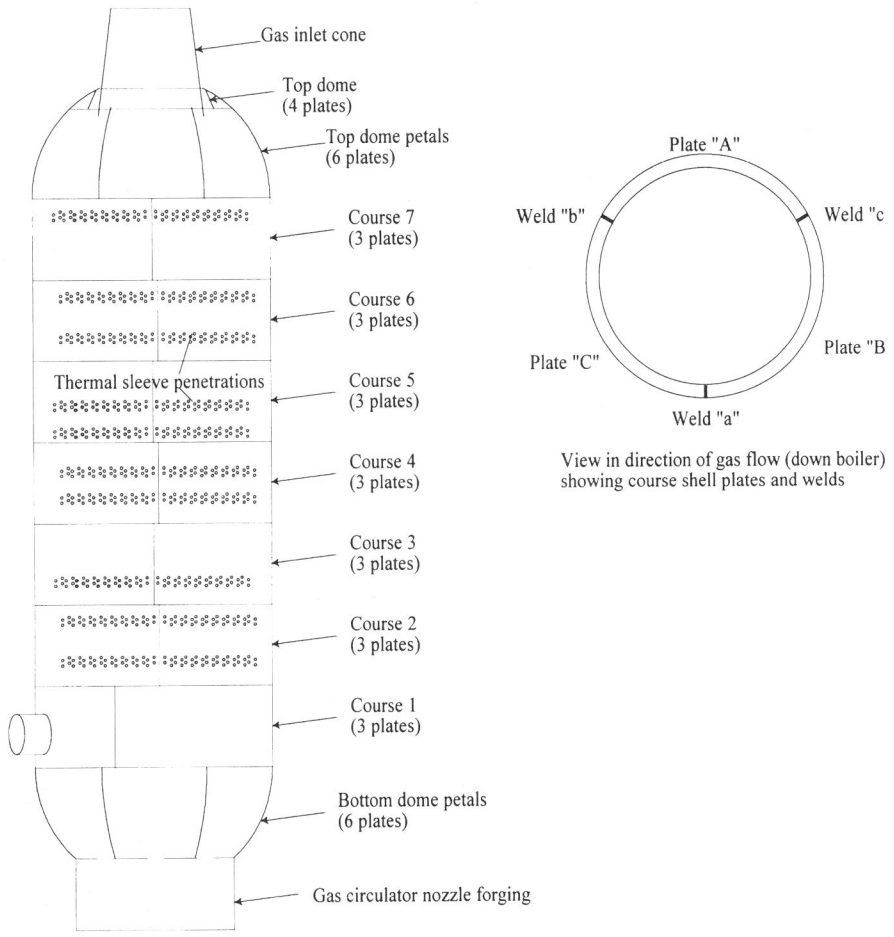

Figure 1: Boiler shell - schematic

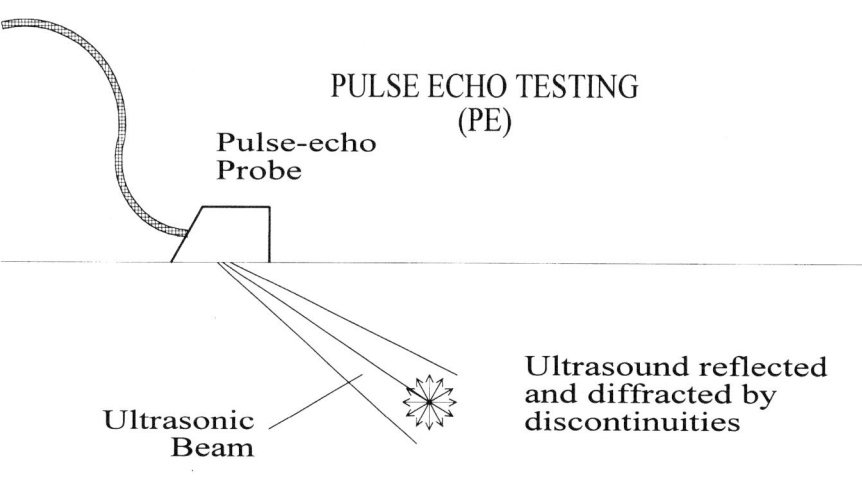

PULSE ECHO TESTING (PE)

Pulse-echo
Probe

Ultrasonic
Beam

Ultrasound reflected
and diffracted by
discontinuities

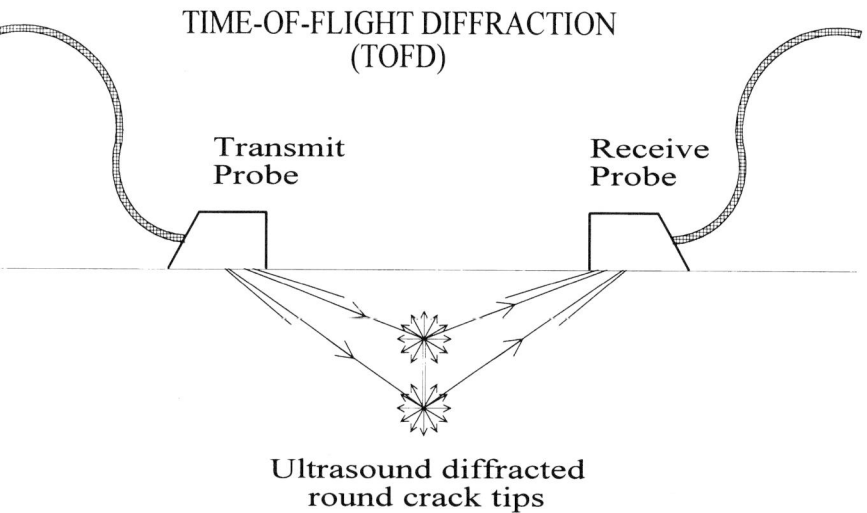

TIME-OF-FLIGHT DIFFRACTION (TOFD)

Transmit
Probe

Receive
Probe

Ultrasound diffracted
round crack tips

Figure 2: Ultrasonic Inspection - Pulse-Echo and Time-of-Flight Diffraction

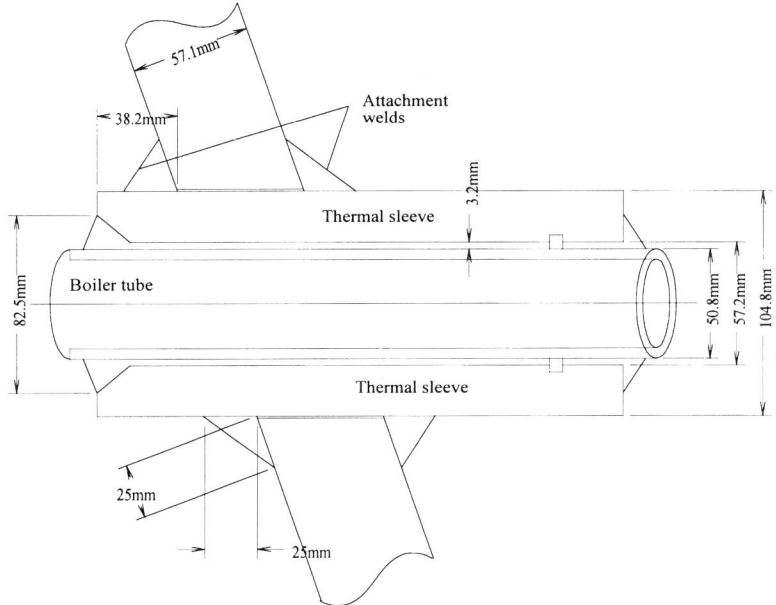

Figure 3: Thermal sleeve with 20° penetration angle through boiler shell

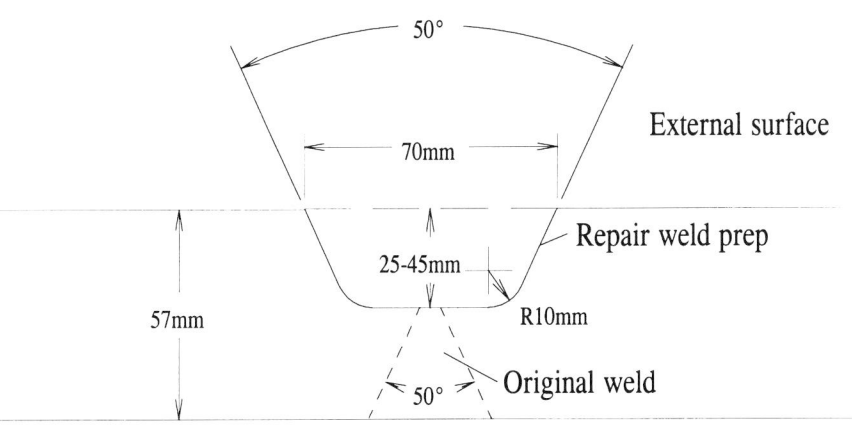

Figure 4: Course 6/7 Repair Weld Prep

VALIDATION CERTIFICATE

AEA Technology
Energy

SIZEWELL A REACTOR 2 BOILERS 2A, 2C and 2D REPAIRS TO WELDS 6/7: VALIDATION OF THE INSPECTION OF THE REPAIR WELD AND ASSOCIATED VOLUME USING THE MANUAL ULTRASONIC NON-DESTRUCTIVE TESTING **PROCEDURE** FOR THE **DETECTION** OF FLAWS

CERTIFICATE NO: VCP 53   ISSUE: SECOND

This is to certify that the inspection process described in the Procedure identified overleaf has been applied by an operator, whose qualifications conform to the requirements of the Procedure, to Test Blocks representative of the repair weld geometry. These Test Blocks contain flaws as described in M/TE/SXA/REP/0058/98 Issue 3. The tests took the form of open trials where the operator had prior knowledge of the defects. The results of these tests, which were carried out under the invigilation of the ME qualification trials team leader and observed on a sample basis by a member of the Independent Qualification Body representing the IVC, taken together with a review of the Technical Justifications and the Qualification Trials Report identified overleaf, show that the Procedure provides for detection of flaws as described in AEAT - 3705 Issue 2.

Day/Month/Year

EXAMINATION PERIOD: 21.10.98 - 23.10.98   Signed (For IVC) ....
Mr C Waites
Head of IVC

EFFECTIVE DATE OF ISSUE: 23.10.98

FOR NOTES ON VALIDITY SEE OVERLEAF | UNEMBOSSED COPIES ARE NOT VALID

ISSUED BY
THE INSPECTION VALIDATION CENTRE
ON BEHALF OF
AEA TECHNOLOGY PLC

AEA Technology plc registered office
329 Harwell, Didcot, Oxfordshire
OX11 0RA. Registered in England
and Wales, number 3095862

Figure 5: Qualification certificate for manual ultrasonic inspection od course 6/7 weld after PWHT

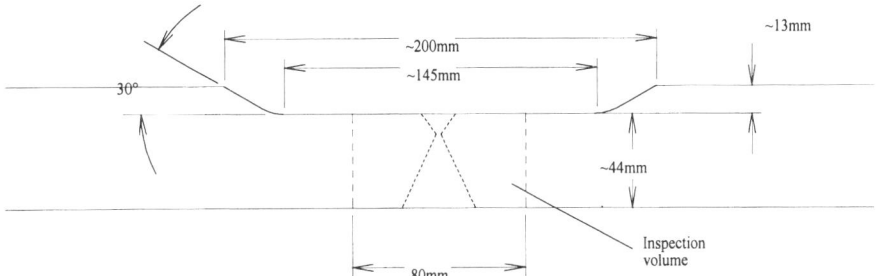

Figure 6: Boiler 2C course 5/6 weld excavation profile (end Elevation)

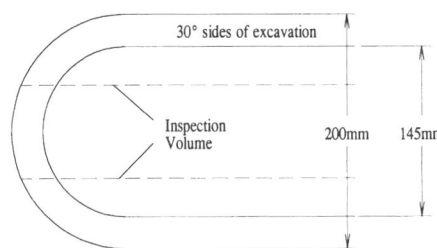

a) Semi-circular end with 30° sides - plan view

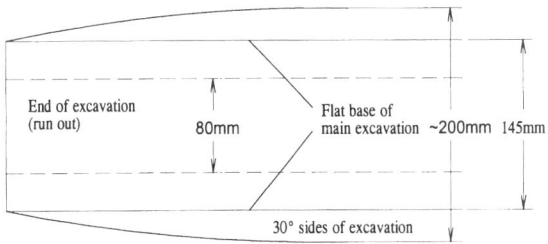

b) Flat end (run out) - plan view

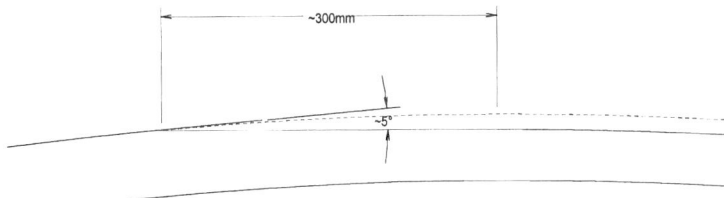

c) Flat end (run out) - side elevation

Figure 7: Boiler 2C course 5/6 weld profile ends (plan and side elevation)

S690/010/99

Project management approach and application to specific technical issues

G S ANDERSON
BNFL Magnox Generation, Berkeley, UK
W J G LITTLE
Mitsui Babcock Energy Limited, Glasgow, UK

1.   ABSTRACT

Having already established that the Sizewell A weld repair project was economic, technically feasible and licensable, the project team were charged with completing the technology development, satisfying the regulator at every stage consistent with the principles agreed at the outset (1), and completing the site repair activities associated with the three defective boilers within a timescale and at a cost consistent with the economic benefits defined in the original scheme paper. These requirements set the scene for an objectives driven process for management and coordination which remained in force until the end of the project. This paper sets out the factors which were considered in the management of the project, the specific challenges which had to be overcome and the way in which these were achieved.

2.   BACKGROUND

The Sizewell A R2 boiler repair project was conceived after exhaustive efforts to examine possible ways of returning the unit to service safely. These options included reduced pressure operation, replacement of the boiler shells, mechanical reinforcements and weld repair with and without stress relief. The only option which could satisfy all safety requirements was a code compliant stress relieved weld repair; the option which was also the most technically challenging. This was reviewed over a 10 week period to examine all aspects of the job, and demonstrate that not only was it technically feasible, but it would be acceptable to the regulators and would be economic. The Magnox Electric Board accepted the positive findings, and granted approval to proceed with the repairs in September 1997.

The team which carried out the feasibility assessment contained the nucleus of the team charged with executing the works, and this minimised familiarity gaps. The team structure and some of the main responsibilities are described in section 3. below. Before discussing this, it is worthwhile briefly discussing the team approach that was intended at the outset. The primary objectives were to "do the right things" and to "do them right".

The first of these was achieved by creating objective driven management roles with full empowerment and accountability within the project, and ensuring that all strategic decisions were made with full cross functional open discussion. Appropriate tools and techniques were used to ensure that the discussions were comprehensive, and that consensus was achieved. Records of how conclusions were reached were distributed to participants and other interested parties. Thus an effective decision making and tracking system was put in place and encouraged throughout the project. This resulted in ownership of the strategy by all involved.

The second of these was achieved by combining well tried and comprehensive QA systems and ensuring that at all levels, the team recognised that these were part of doing the job, not an add-on. Indeed work could not be carried out without all aspects of the quality system being in place first. For example, work could not proceed at the work face without a work control card, which in turn could not be authorised until all supporting quality paperwork was in place, e.g., quality plans, method statements and necessary safety clearances such as Independent Nuclear Safety Assessment (INSA) approvals. This INSA requirement is an important feature of the way work is carried out within the company. It ensured that work could not proceed on the plant until all safety and integrity related clearances were in place. This had the dual advantage of minimising abortive work, and ensuring that detailed safety or integrity issues were progressively cleared as the job progressed. This helped significantly in making the final safety case.

The culture of closing out issues as the job proceeded, even if they were not on the critical path, ensured that the final hand-over of the boilers back to the station was quick and painless. An example of this was the collation of the lifetime records. This is an activity which can extend for several months, especially at the end of a long or technically complex project. By assembling records and reports as the work progressed, this activity took only two weeks after the finish of the project to complete, and did not delay plant start-up.

3. TEAM STRUCTURE AND COORDINATION

As well as requiring significant site engineering support, the project required specialists in fields which included material properties, structural integrity, welding, heat treatment, safety case preparation and Non Destructive Testing (NDT). These specialists came mainly from within Magnox Electric and from the prime contractor, Mitsui Babcock Energy Services Ltd (MBESL). Support was also acquired from other contractors and agencies as required.

The MBESL site team was responsible for controlling the site implementation engineering. In this, they were supported by Magnox station engineers who provided detailed plant knowledge. Technical support was provided by Mitsui Babcock Energy Ltd (MBEL) engineers from Mitsui Babcock Technology Centre in Renfrew and Magnox engineers from Berkeley Centre who provided design and safety case compliance support. The design engineering was provided by Mitsui Babcock Technology Centre. They controlled the mechanical engineering associated with, e.g., the duct splitting, the welding, the lifting of the boiler tube banks and the profiled repair of boiler 2C. Welding development and welder qualification were carried out by MBESL staff from the Welding Department at Tipton.

Magnox engineers from Technology and Central Engineering (Berkeley Centre) supported the MBEL welding design, but took the lead in materials issues and the development of the heat treatment. The NDT development and support was also carried out at Berkeley, as was the development of the overall safety case.

Geographically, team members were spread throughout the UK. It was crucial to the timely completion of the project that different company cultures and locations did not compromise the performance of the team. Early integration of individuals into multi-company teams with clearly defined sub-objectives, and effective communications systems ensured that efficiency and motivation were maintained and nothing was missed. Modern communication methods such as E-mail and video-conferencing were extensively employed.

The bulk of the work was managed between two large teams, each with its own manager. The first was the Site Implementation Team, whose role was to ensure that site work progressed swiftly in accordance with the technical, safety case and site license requirements. This team was also charged with setting up and maintaining safe working conditions. The second was the Technical Team, whose role was to provide design and technical input and interim safety clearances to enable the Site Implementation Team to perform safely and enable site works to proceed. They also were responsible for constructing a final safety case which would eventually enable a return to power. Both of the teams were further divided into working groups to address respective sub-issues which are discussed later.

All emerging major issues, together with overall progress, were reviewed weekly at Lead Team meetings which brought representatives from all parts of the project together, as well as from the station production department. Links to appropriate sites were made by video conferencing. The objective of this team was to keep the project focused on doing the right things correctly, managing risks and ensuring nothing was overlooked. Conclusions reached cascaded to other teams and individuals in a controlled manner and actions tracked. Emphasis changed during the project. During the early stages, the main items discussed related to defining the detailed workscope, the formation of the team, getting work started (including preparation of stage submissions and method statements) and working practices. Later, emphasis was on bulk liquidation of the work, and the removal of obstacles from whatever source (resources, licensing, technology, etc.). Finally, emphasis was placed on project closure. Appropriate indicators were monitored at each stage.

4. PROJECT MILESTONES AND THE CLEARANCE PROCESS

Project milestones were reflected in part by hold points imposed by the HSE regulator, the Nuclear Installations Inspectorate (NII). More detailed Hold Points were imposed by Health, Safety and Environmental Department (HSED), who are Magnox's internal regulator. Each hold point required demonstrations that the proposed work would satisfy all necessary safety requirements and that necessary quality arrangements were in place. This was achieved through a variety of documentation. In particular, use was made of Stage Submissions which were pre-defined sub-divisions of the final safety case, and which were approved internally by HSED as part of the continuous safety case clearance process. This approach had been agreed in advance by the Company's Nuclear Safety Committee (NSC) and the NII.

The requirements for each hold point had been carefully considered to ensure that it was clearly demonstrated that all work conceived up to the following hold point would not compromise a successful repair. In this way, the hold points created essential quality checks for both the project and the regulators.

The following paragraphs relate to the clearance of NII hold points only. A similar process for securing written clearance was also applied to pass HSED hold points.

Clearance from the NII was required before any work was allowed to commence and this constituted the first hold point. They required evidence that the project was organised adequately to carry out such repairs. This was provided in the first of the stage submissions (2). This document contained the top tier quality arrangements and formed the basis for detailed process procedures which followed later.

The first NII hold point also allowed the machining of the boiler shell to the agreed weld preparations to proceed. In line with the above guidelines, it was necessary to show that the weld preparations designs were well thought through, and would satisfy the requirements of later activities. The overall shape was justified by providing a generic stage submission describing the temper bead welding technique and demonstrating that the shape was optimised to that weld process. The preparations designs had to be optimised in length, depth and width and also in overall profile.

The weld length was determined by the results from material sampling down to approximately 20mm to ensure that the ligament beneath the proposed repair was sound. This process was referred to as "boat" sampling. It became clear during the boat sampling programme that the original weld Heat Affected Zone (HAZ) close to the outer surface of the boilers throughout the circumference of each of the defective welds contained significant creep damage. A project decision was then taken to excavate the complete circumference of each 6/7 seam weld to prevent possible new cracking during the heat treatment. The justification for complete circumferential excavation was provided by stage submissions describing the test results.

The depth of the weld preparation was determined by the extent of the creep damage in the samples and the requirement to excavate pre-existing defects and any associated local creep damage material from the remaining ligament. These deeper excavations were mainly local to the main defects. Justification was also provided where it was decided not to excavate minor buried defects from the ligament. Where there were no local defects requiring to be removed, an overall depth of 25mm ensured that all creep damaged material was eliminated.

The width was pre-determined by the need to completely remove the HAZ of the original weld, with local area increases in width necessitated by defects in some intersecting axial welds.

Finally, the overall shape was justified by providing a generic stage submission describing the temper bead welding technique, and demonstrating that the shape was optimised to that weld process.

There were no NII hold points specifically against the 2C seam 5/6 profiled repair as such, although HSED had included this in their schedule, and the NII were kept fully aware of developments as the final profile design developed. The integrity of the 5/6 repair however needed to be confirmed prior to pressurisation, see hold point 3 below.

The second NII hold point was subdivided into three, one for each boiler, which had to be cleared prior to welding. The start of each weld initiated a cascading series of linked and unstoppable events which ended with the heat treatment. It was necessary to complete each of these activities as quickly as practicable, as the weld area temperatures would not be permitted to drop below about $150°C$ until after the heat treatment because of the higher risk of cracking. Thus virtually all issues associated with the welded repairs and heat treatment had to be concluded prior to proceeding beyond hold point 2. Again these were largely incorporated in the stage submissions.

The third hold point had to be cleared before the boilers were allowed to be pressurised. This required information which demonstrated that the welding and heat treatment had been correctly implemented, the boilers had been correctly re-assembled including the lowering of the boiler tube banks and the re-connection of the reactor outlet gas ducts. Much of this confirmatory information was provided by plant completion statements put together by the Independent Third Party Inspection Agency (ITPIA) whose role is discussed in 5. below. These statements were in turn justified by completed quality plans. It was necessary to demonstrate for the 2C 5/6 profiled repair that the remaining ligament and the overall shape were ASME compliant, and also that the profile would lend itself to meaningful NDT. Confirmation of the integrity of the repairs, including the 2C 5/6 repair, and the adjacent welds influenced by the heat treatment, was required and this was provided by NDT reports.

Clearance of the fourth hold point was necessary (but not sufficient) to allow the circuits to return to nuclear heating. This required satisfactory results from the monitoring arrangements in place (mainly strain gauges on the 2C 5/6 profiled area) during the non nuclear pressurisation up to 150 psi. Consent to go to nuclear heating was granted to the station following satisfactory resolution of a number overall outage issues, of which the repair of the boilers was but one, albeit significant, issue. This enabled the reactor to operate in "commissioning mode".

The fifth and last hold point was cleared following successful monitoring of the boilers during the preliminary period of operation. This enabled the reactor to proceed to normal operation.

Documentation supporting the designs for each boiler were subjected to detailed review by the independent assessors (described in 5. below) and their findings submitted to the regulators as part of the package for each hold point.

These hold points provided clear management objectives which were used to drive the project.

5. INDEPENDENT ASSESSMENT

Assessment of safety submissions is managed internally within Magnox using an Independent Nuclear Safety Assessment (INSA) team whose role is to agree, qualify or reject submissions before they are passed on formally to the NII. The technical complexity of the work and its key safety significance called for a high degree of assurance that the cases were sound. The INSA team, project team, and the regulators (internal and external) therefore sought assurance that proposed activities were suitable for purpose, and that they were carried out correctly. Confidence was provided by means of independent assessors. A framework within which the work was to be performed was provided by ASME XI, and this was used to assist the assessors.

The verification that the agreed processes were code compliant was provided by the Independent Design Authority (IDA), Royal and Sun Alliance. They achieved this by careful review of each stage submission or supporting document related to the design, and assessing the information presented against the requirements of the ASME XI code. Further design confidence was provided by peer review of appropriate documentation by experts from the University of Manchester Institute of Science and Technology (UMIST) and the Welding Institute (TWI).

The Independent Third Party Inspection Agency (ITPIA), Kennedy and Donkin, ensured that the project processes were carried out correctly. The ITPIA role for this type of project is an ASME requirement.

The Non Destructive Testing (NDT) aspects of the project were reviewed and qualified as required and this was reviewed by an Independent Qualification Body (IQB) in accordance with the European Network for Inspection Qualification (ENIQ) requirements. The organisation selected for this role was the Inspection Validation Centre (IVC),who are part of AEA Technology. All NDT processes underwent scrutiny, including detailed qualification of the more critical inspections.

Rigorous controls were imposed to ensure that all relevant aspects of the project, including change control, came under the scrutiny of these bodies as appropriate. This provided added comfort to the regulators, and helped make the clearance process smoother.

6. TECHNICAL ISSUES ARISING DURING THE PROJECT

During the course of the work, many issues arose where the correct solution was not the same as that originally envisaged. Although a rigorous risk assessment was carried out to support the scheme proposal prior to sanction, and refreshed during the project, some issues materialised which had been assessed as having a low probability of occurrence, and some emerged which were not even foreseen during the risk assessments. This section discusses some of these issues to illustrate how the team adapted to these new challenges.

The first issue discussed here is the way in which the project team dealt with the discovery during preliminary destructive examinations that the HAZ material circumferentially remote from the defect areas still exhibited significant creep damage rendering it probably unfit to undergo a further heat treatment in its current condition.

The original plan was to excavate just beyond the defects lengths involving 8M of welded repair. If the entire circumference of each boiler needed excavation this would involve 66M of welded repair.

The original strategy covered the sequential welding of one boiler at a time. This optimised the programme time for welding against the additional labour and equipment costs involved in parallel working.

Analysis of the original strategy with the increased welding times resulted in an unacceptable contract extension and loss of generation costs. Parallel working on several boilers became a necessary consideration.

Key considerations to be taken into account were:

- the feasibility of parallel welding.

- cost of additional heat treatment equipment (transformers, heating pads, controllers, recorders, cables, etc.), needed to enable simultaneous welding pre-heat and heat treatment.

- optimum sequencing of boilers.

- transport arrangements for moving equipment from boiler to.

- the likely generation benefit from a shorter programme.

To resolve this complex issue quickly and to ensure team sign-on, a meeting of responsible delegates from MBESL, Magnox and Didcot Heat Treatment (contractor) was held to carry out cost benefit analyses of all options. The option finally agreed was to purchase sufficient equipment to enable two boilers to be repaired simultaneously.

One interesting aspect was the realisation that equipment had to be moved twice, from the 2D/C boiler house (to repair 2D) to the 2A/B boilerhouse (to repair 2A), then back again to

2D/C boilerhouse to repair 2C. This double movement resulted from the belief that congestion resulting from two simultaneous repairs in one boiler house would have led to impossible working conditions. Experience of working conditions during the repair demonstrated that this was a wise decision.

Once the recommendation emerged from the team, advice was referred up rapidly through the accountability chain on cost implications and the reasons for them to enable additional money to be released.

Other areas which proved to be more complicated than initially perceived were the heat treatment process and the raising of the superheater tube bank off its support pad to reduce pad weld stress during heat treatment.

The heat treatment specification became more demanding once it was realised that only very benign thermal stresses could be tolerated during the heat treatment, and there was a need to ramp very rapidly across a temperature band (600-650°C). Also tests carried out on a full scale rig indicated that the insulation needed to be very comprehensive to maintain an acceptable temperature distribution. To ensure all aspects of heat treatment development and application were effectively managed and requirements accurately communicated to site, all heat treatment issues were treated as a project in its own right with its own project manager from Magnox reporting to the lead team. This project team controlled the development work and set the allowable tolerances on temperatures during the heat treatment. Two teams consisting of specialists with metallurgical, stress analysis and heat treatment backgrounds were formed to supervise the heat treatment over the required 36 hour cycle. The close attention to detail resulted in very few thermocouples being outwith their specified band. The review process was completed in a few days to confirm that the target criteria for the heat treatment had been met. The personnel in the teams were maintained for the heat treatment of each boiler to ensure experience gained could be applied to the next unit.

The complexity of raising the 100 tonne superheater tube banks in cramped working conditions from their support pads and suspending them from the boiler inlet cone such that they could be successfully reinstated afterwards was similarly much more demanding than expected at the outset. This was also treated as a stand-alone project with its own project manager, this time from MBEL Technology Centre in Renfrew. The key to the success of this activity was the preparatory work. A period of several weeks was spent proof testing components or large parts of the lifting frame and carrying out inspections of the major lifting components. A team of engineers with experience in lifting of structures, instrumentation and stress analysis was formed. The team inside the boiler lifted the tube bank incrementally, relaying deflections to the structural analysts who had set up the design model for the frame and tube bank outside the boiler. Deflections and strains were fed into this model on a continual basis during the lifts to ensure that even loading of the frame and tube bank was occurring. This detailed assessment ensured the lifts proceeded quickly and safely, the lifting of the bank by 15 mm taking a period of approximately four hours. The care taken to ensure a balanced and even lift paid dividends in that the lowering of the tube bank back to its original position on the support feet was achieved within a few hours.

Many other issues arose and were addressed in a similar fashion throughout the project.

7. STAKEHOLDER MANAGEMENT

During the feasibility assessment and during the project itself, good management of stakeholders was important to maintain support for the project and facilitate rapid decision making. Two of the main stakeholders in this instance were the Magnox senior management and the regulators (internal and external). The objective of the stakeholder management was to inspire and retain confidence that the project was truly under control and that decisions taken were consistent with stakeholder requirements (licensing, safety or business plan). It was intended that such confidence would lead to shorter stakeholder deliberation time when issues arose.

The requirement for good stakeholder management was identified at the initial project risk assessment. Countermeasures were identified, which were applied throughout the project, and included regular, accurate and relevant communications. Wherever possible, requirements from the project were agreed in advance, and promised information met agreed timescales, particularly with licensing submissions.

Practical manifestations of these countermeasures in the licensing area appeared, for example, in the hold point schedules, where summaries of information which would be provided to clear the hold points were agreed up front with the NII, eliminating the need to negotiate the extent of information to be provided during the assessment period. Furthermore, regular (at least weekly) project verbal updates were provided to the NII to ensure they had a clear up to date picture of progress and early warning of developing issues. These were additional to periodic NII meetings to discuss technical issues and general progress. Advance good quality drafts were sent where appropriate to speed the licensing assessment process, and clear indications of which licensing clearances were most critical were given to enable the NII to prioritise effectively. Finally regular feedback from the ITPIA and independent assessors added to the confidence that the right things were being done properly. These approaches inspired trust, and probably resulted in significant reduced assessment time.

In the case of Magnox senior management, information on cost, programme, progress and licensing was submitted in progress reports, approximately monthly, with any significant emerging issues being flagged up as they arose. An accurate and early explanation of the impact of these issues was made with no attempt made to paint a rosy picture. Again this inspired trust and confidence, and resulted in executive decisions being made quicker than might otherwise have been the case.

8. PROJECT CLOSE-OUT

As work approached the end, the project emphasis swung from monitoring work progress to the close out of outstanding issues. These included signed off quality plans (which incorporated a number of other quality documents such as work completion statements), closed out technical queries and non conformances, lifetime records and outstanding licensing concerns. The latter two items were recognised as posing significant risks to the wind up of the project and to start up date if they were not well managed, and so they were addressed continuously throughout the project. Once a work area was completed, lifetime records were compiled according to ASME requirements without delay. In practice, the final project lifetime record was in place one week after completion of the physical work. Licensing issues were also closed out as they arose. The tidying up of licensing issues as the job progressed was also helped to some extent by the licensing hold points. Thus continuous monitoring and management of close out items by the lead team, particularly during the final two months of the project, led to an uneventful and predictable end to the project.

9   LESSONS LEARNT

Shortly before the end of the project, a series of 8 short (2 hour) workshops were held with appropriate team members to review performance and make recommendations for improving performance of future projects. Topics were:

1   Evaluation of viability of Project

2   Organisation of Site Activities

3   Organisation of Technical Activities

4   Planning, Interfaces and Control

5   Project Procedures and QA

6   Licensing Arrangements

7   Scheme Financing and Control

8   Contracting Strategy

The agenda for each workshop was typically:

1   Steps carried out by Project

2   What were the good points?

3   What could we have done better?

4   What should we have done instead?

5   What would that make the process look like?

6   Action plan to achieve benefits

The information derived from these workshops was recorded in concise (~3 page) stand alone reports and distributed to participating organisations for future reference. The information was also incorporated into the final project close-out report. Such findings may be useful for future project managers to enable them to avoid pitfalls. The breadth and thoroughness of the process makes it difficult to record the findings here. However, a sample of countermeasures to project shortcomings under each of the topics is given below:

1.   Develop project management culture within Technology, majoring on technical and licensing issues.

2.   Make Sizewell A project Quality Arrangements documentation available to future project managers as basis for their quality arrangements.

3.   Closer understanding at early stages of the job between submission authors and site personnel of each others' requirements.

4.   Guidance from project manager on the structure and formatting of the project programme to make it generally more understandable to all.

5.   Early audits to be put in place to assist rather than catch out.

6.   Early dialogue with NII to clearly identify go / no-go areas

7.   More regular project and commercial awareness briefings to team members.

8.   Client planner to get more directly involved in contractor planning.

10.   ACKNOWLEDGEMENT

We would like to thank our many colleagues at the Berkeley Centre and at site who have contributed to this work, particularly Mr I Lingham (Morson). This paper is published with the permission of the Director Technology and Central Engineering, BNFL Magnox Generation.

11.   REFERENCES

1.   P J Jeans, NUCLEAR SAFETY COMMITTEE. Sizewell A Power Station. Paper of Principle for: 1) The Development and Implementation of a Weld Repair Procedure and Post Weld Heat Treatment for the Defective Course 6/7

Circumferential Seam Welds in Boiler Shells 2A, 2C and 2D; 2) Profiling of the Course 5/6 Defective Weld in Boiler Shell 2C; 3) The Development of a Safety Case for Return to Service of Reactor 2. NP/SC 4436,   September 1997

2.   J Field, Sizewell A Reactor 2 Boiler Repair Project: Project Implementation Arrangements. SIZA-R2-BRP/SS1 Issue 4,. May 1998.

Inspection qualification

V M BABOROVSKY
BNFL Magnox Generation, Berkeley, UK

ABSTRACT

Routine inspections of Sizewell 'A' Reactor 2 boilers revealed significant cracking in four circumferential welds and these have been repaired. Since there was no pressure test on completion of the repair, great emphasis was placed on NDT inspections. The details of the inspections are covered in a separate paper(1). To demonstrate and confirm their capability, several inspections were qualified following the ENIQ (European Network for Inspection Qualification) methodology. This allows for a graded approach to qualification, the rigour depending on the importance of the particular inspection, its novelty and its difficulty.

Four main inspections were subject to full qualification, involving technical justifications, practical trials and an Independent Qualification Body (IQB). The paper concentrates on these four qualifications and discusses the main lessons learned. The problems encountered were all overcome and the adequacy of the inspections successfully demonstrated.

1.   INTRODUCTION

Inspection qualification provides confidence that an inspection has the capability to achieve the required or claimed performance, the evidence for inspection capability being reviewed by a third party. Other quality assurance (QA) measures need to be in place to ensure that the qualified inspections are implemented correctly and consistently.

The concept of inspection qualification is not new. Most notably, some ten years ago, an extensive programme of inspection qualification, which at that time was called validation, was carried out by one of the predecessors of BNFL Magnox, the CEGB, for the Sizewell 'B' pressurised water reactor (PWR). Since then, inspection qualification has also been incorporated in the ASME pressure vessel code, as "performance demonstration" in Appendix VIII to ASME XI(2). More recently, the European Network for Inspection Qualification (ENIQ) has developed a methodology, published in 1997(3), which is quickly gaining wide international acceptance by both utilities and national regulators, including the Nuclear Installations Inspectorate (NII).

The qualifications described in this paper were the first ones in BNFL Magnox Generation to be carried out according to the ENIQ methodology. The paper explains the reasons for choosing ENIQ and the experience with this new methodology. The details of the inspections themselves are covered in a separate paper(1).

2. CHOICE OF APPROACH

The repair of the Sizewell 'A' boiler welds was required to be compliant with the ASME code. Whilst ASME does not require NDT qualification for boiler weld repairs, since a post weld repair pressure test was not feasible, great emphasis was placed on NDT inspections, and it was agreed with the NII to qualify certain main inspections.

The approach used for the validations for Sizewell 'B' involved the preparation of technical justifications (TJs) for a number of different ultrasonic inspections and relied heavily on blind trials. The philosophy was to justify the inspection capability for all plausible defects and then to demonstrate the capability on selected defects implanted in test assemblies of the correct geometry. The data interpretation engineers were required to obtain individual certificates which were based on their performance in the blind trials.

In most cases, the Sizewell 'B' validation process was conducted by an external, independent validation body. Valuable lessons were learned from the exercise and many of these were later adopted in the ENIQ methodology. For example, ENIQ is careful to separate procedure qualification from personnel qualification, which was not the case for the original Sizewell 'B' validations, because it could be difficult to disentangle operator error from a shortcoming in the procedure.

The ASME approach to qualification is extremely prescriptive regarding components, defects and acceptance criteria. It does not address boiler seam welds, and hence it cannot be applied directly to the boiler repairs. The ASME approach relies on blind trials and no use is made of evidence from other inspections, trials, development work or arguments based on theoretical modelling. Furthermore the European Regulators, in reviewing ASME XI Appendix VIII, note that, "the purpose of the chosen detection/false call acceptance criteria is not to provide evidence that an ultrasonic system has a certain performance ..... but rather achieve a reasonably effective discrimination between capable and less capable ultrasonic systems"(4).

The ENIQ approach, being the most recent, incorporates experience from the earlier validations and qualifications and owes much to the Sizewell 'B' validations. With several new guidance documents issued since the publication of the main methodology document(3), the ENIQ process is now well documented.

Unlike ASME, the ENIQ methodology is very flexible, with much left to agreement between the utility and the regulator. It encourages the use of TJs and allows for a graded approach to qualification, ie the rigour of the qualification depends on the safety significance, difficulty and novelty of the inspection. Thus ENIQ recognises that requirements for inspection qualification can range from none for routine inspections, to a TJ with open or even blind trials for critical complex and novel inspections. ENIQ draws a distinction between equipment/procedure qualification and personnel qualification: while blind trials may be required for personnel qualification, open trials are regarded as appropriate for equipment/procedure qualifications. Thus ENIQ encourages the use of evidence and argument in TJs backed up by appropriate trials and should therefore be the most effective of the three approaches. Since it does not rely on extensive practical trials, this has the further benefit of reducing the requirements for expensive test specimens and for lengthy trials and this should enable qualification to be obtained more speedily.

The ENIQ approach was adopted and this for a number of reasons: it is soundly based, well thought out and intellectually the most appealing, it represents the best international practice and it should also be the least likely to cause delays to the repair programme.

Thus the qualification involved preparing technical justifications and practical trials. Open trials were used to show that the appropriate procedures, equipment and personnel can reliably detect and correctly report a range of representative defects contained in test blocks specifically designed for the qualification.

3.   SIZEWELL 'A' PLANT AND DEFECT DETAILS

There are two reactors at Sizewell 'A' and each has four boilers, all manufactured from a low alloy ferritic Mn-Cr-Mo-V steel. The boiler vessels are upright cylinders, with domes at the top and the bottom. The cylindrical part of each boiler is made up of seven 57 mm thick courses, with three 120° segments in each course, as illustrated in Figure 1. The circumferential welds between the courses are about 22 m long while the axial welds between the segments are between 2.1 and 3.2 m long.

In-service inspections found cracks in the Course 6/7 welds of three boilers, 2A, 2C and 2D. A further crack was found in boiler 2C, Course 5/6 weld. These welds are of the double-V prep design as illustrated in Figure 2. The defects were surface breaking and were identified as stress relief cracking(5).

The best estimates of the maximum depths of the cracks in the course 6/7 welds ranged from 22 mm in boiler 2D to 33 mm in boiler 2A. For the three course 6/7 welds, the excavations extended all the way round the boiler with the weld preparation shown in Figure 3. The cracking in the Course 5/6 weld was less deep (~13 mm) and was repaired by excavation and re-profiling of the adjacent area to reduce stresses to an acceptable level and to enable high quality NDT. The profile is illustrated in Figure 4.

4.   INSPECTIONS QUALIFIED

One of the ASME XI requirements is that repairs are inspected with the same techniques as those that detected the original defects. Thus MPI and manual ultrasonics, supported by automated pulse echo and time-of-flight diffraction (TOFD) ultrasonics, were to be used.

The weld repair programme as a whole required a large number of inspections, some for reasons of structural integrity and safety and some because of commercial considerations. The integrity considerations meant that in addition to the repaired welds, inspections covered also a number of adjacent welds. Further inspections were carried out at various stages of the repair process for commercial reasons, to avoid costly delays which could occur if defects had to be removed at a later stage, following the final structural integrity inspections.

With such a wide range of inspections, the graded approach to qualification, permitted by the ENIQ methodology, has proved to be particularly useful and advantageous. The inspections were divided into three categories according to their safety importance, novelty and difficulty. The three categories and the appropriate agreed levels of qualification are indicated below:

1. Routine inspections — Standard BNFL Magnox Generation procedure
   Personnel with relevant certification

2. Intermediate inspections — Capability statement for the procedure
   Personnel with relevant certification

3. Novel, and/or safety critical inspections — Procedure/equipment: Technical Justification (TJ)
   Open trials
   Personnel: Relevant certification
   Additional exams/certification
   Open or blind trials

The purposes of each of the principle sources of evidence for the qualifications are as follows:

Technical Justifications provide evidence to show that the inherent capability of the inspection is adequate. They include physical reasoning, mathematical modelling and results of development work.

Open trials demonstrate that techniques applied according to the procedure, by qualified operators, achieve the required performance.

Blind trials are complementary to TJs and open trials. They provide added confidence in operator performance.

Capability statements summarise evidence on inspection capability, for a number of situations.

Five capability statements were produced, covering defect sizing, manual ultrasonic inspections of the axial welds and manual ultrasonic and MPI inspections of relatively short welds associated with fittings close to the repaired welds. The statements were all endorsed by two independent bodies, Rolls-Royce Marine Power and Royal Sunalliance.

The paper will now address inspections for which full qualification was specified, ie requiring a TJ, trials and certification by an independent qualification body (IQB). There were four such inspections and their details are given in Table 1 below.

It should be noted that all of these were ultrasonic inspections of the circumferential weld repairs. The aim of these inspections was either defect detection or, in Case 2 only, detection and measurement of defect growth during stress relief. Blind trials were only required for Case 2. This was a complex and novel inspection designed to determine whether there had been any stress relief cracking resulting from the post weld heat treatment (PWHT). Inspection data was collected before and after PWHT and compared to assess whether there had been any significant change. This qualification proved more difficult than those for the defect detection inspections.

Table 1. Inspections requiring qualification

No.	Inspection Region	Repair Stage	Scope	Purpose of Inspection
1	Boilers 2A, 2C & 2D: course 6/7 welds	After excavation	Procedure qualification for defect detection	Commercial: to ensure the removal of all significant defects prior to welding
2	Boilers 2A, 2C & 2D: course 6/7 welds	After repair welding. Data acquired before and after post weld heat treatment (PWHT) at ~160°C	Procedure and data analyst qualification for monitoring for defect growth by comparison of inspection data from pre and post PWHT	Safety: to determine whether PWHT has caused defects to grow or initiate
3	Boilers 2A, 2C & 2D: course 6/7 welds	Post PWHT (at ambient temperature)	Procedure qualification for defect detection	Safety: to detect defects exceeding the specified acceptance criteria and to trigger growth monitoring assessment
4	Boiler 2C: course 5/6 weld	After repair excavation	Procedure qualification for defect detection	Safety: to detect defects exceeding the specified acceptance criteria

Each of the four qualifications is described briefly in Section 6.

5. THE QUALIFICATION PROCESS

In accordance with the ENIQ methodology, one team, the inspection development team led by BNFL Magnox Generation personnel, was responsible for assembling the evidence for the qualification, including the preparation of the inspection procedures and technical justifications. The qualification was implemented by a separate independent team, composed of both BNFL Magnox Generation and external members. The activities were overseen by an Independent Qualification Body (IQB), constituted by the Inspection Validation Centre (IVC) of AEA Technology and staffed by personnel with considerable experience of inspection validation and qualification. The IQB, upon being satisfied with the qualification, issued Validation Certificates for the different procedures and, for the qualification including blind trials, also additional certificates for the individual data analysts.

The structure originally envisaged comprised a qualification team, a blind trials team and the Independent Qualification Body.

The qualification team was responsible for the qualification procedures and managing the qualification process as well as for ensuring that inspection personnel met the necessary requirements, including certification and familiarisation with the procedures and equipment.

The blind trial team was specifically responsible for the blind trials, specifying the blocks and their content and preparing the blind trial data sets for the trials. The details of the test blocks and of the data had to be kept secure and confidential so as not to compromise the blind trials.

The IQB commented on the qualification documentation, witnessed the trials and audited the arrangements for ensuring that the inspectors had the required qualifications, training and experience. When satisfied, the IQB endorsed the qualification and issued the qualification certificates.

The qualification documentation required by ENIQ includes inspection procedures, qualification procedures, TJs of the inspection procedures, TJs of the test blocks and defects used in the trials, reports on the results of practical trials (both open and blind) and the qualification dossier.

In view of limited time scales, the original qualification procedure had aimed to provide a good flow of information, and hence minimise delays, by producing a range of interim documents that would provide all parties with information and evidence in advance of the formal issue of the technical justifications and other key documentation required by ENIQ. However, the production of these documents and commenting on them took effort away from the production of the key documents and proved to be counterproductive. Consequently, following the first qualification, this approach was dropped and the information flow was ensured by regular progress meetings between the qualification teams, the IQB and the inspection development team.

The meetings also led to improved cooperation between the qualification teams and the IQB, which then worked jointly as a single team, rather than in an iterative way. This accelerated the qualification process as it helped to overcome the problems that arose much more speedily.

6.   QUALIFICATION OF THE INSPECTIONS

Significant issues from the four qualifications listed in Table 1 above are described below.

6.1 Course 6/7 Repair Weld Profile After Excavation and Before Welding
This was the first qualification carried out. There was more extensive commenting and resolving of issues arising from the TJ than had originally been anticipated and, as described above, some modifications to the qualification strategy were introduced based on this experience. However, the problems encountered were all resolved and the open trials and qualification proceeded to a satisfactory conclusion, with qualification certificates issued for both the manual and the automated inspections. The certificate for the automated inspection is reproduced in Figure 5.

One problem arose after the qualification certificates had been issued. For two welds, the excavation had to be locally widened. The procedure and technical justification had not allowed for this possibility and an addendum to the technical justification was produced which demonstrated that the inspection still had the required capability for the widened regions, provided they met certain specified criteria. This addendum was endorsed by the

IQB, and all the widened excavations met the specified criteria.

6.2 Course 6/7 Repair Weld - Monitoring for Defect Growth

This inspection was the most novel and complex. The objective was to compare inspection data collected before and after PWHT to determine whether there had been any significant change arising from the PWHT.

One approach would have been to assess the accuracy of defect sizing and then to assess the capability to detect change by combining the errors from the two inspections. However, BNFL Magnox Generation felt that in most cases this would be unduly pessimistic since, if the inspections were highly repeatable, a more direct comparison of the data would yield better capability.

The inspection data comprised both pulse-echo data, for the volume close to scanning surface, and TOFD data for the rest of the thickness. The best measurement of change could be achieved if the same technique could be used both before and after PWHT since certain systematic errors could then be eliminated. However, neither of the two techniques covered the full thickness range and rules had to be devised to cater for a range of possible situations, allowing for possible defect growth and absence of usable data from one of the techniques either before or after PWHT. This complexity was reflected in the qualification and the criteria for qualification took some time to finalise.

The qualification was implemented in two parts. In the first part, the data acquisition procedures were qualified by TJs and open trials. This qualification resulted in additional data quality checks being introduced for the TOFD data.

In the second part, the data analysis procedure was qualified separately, and included the blind trials in which individual data analysts were each presented with data from a number of defect pairs and had to report whether growth occurred or not. The review of the technical justification and the procedure led to a number of changes to the procedure to make it more rigorous. In particular, strict and comprehensive rules were defined as to which technique (pulse-echo or TOFD) should be used under particular circumstances. As a part of the trials, the data analysts had to pass a written examination to ensure that they fully understood these rules. Five out of nine candidate data analysts were qualified and a certificate is reproduced in Figure 6.

As could be expected with a technically novel procedure, the qualification encountered several problems. However, a way was found through all these difficulties and the qualification certificates were issued by the time they were required by the repair programme.

With nine data analysts involved, this qualification produced a considerable amount of information on defect sizing and it was agreed that the sizing accuracy claims made for these inspections could in part be based on the results of the blind trials.

6.3 Course 6/7 Repair Weld After PWHT

By contrast, this qualification proved to be straightforward and proceeded with no significant problems.

6.4 Boiler 2C Course 5/6 Weld After Excavation

This qualification was again fairly straightforward. The principal technical problem was that there was significant mismatch of the course 5 and course 6 plates at the inside surface. As a consequence, credit could not be taken for detection of defects by sound beams reflected off the inside surface, because they could be affected by the mismatch. However, it was demonstrated that the inspection did have the required capability even without the reflected beams and the inspection was qualified.

7.   VALUE ADDED AND LESSONS LEARNED

General. The discipline of preparing a technical justification helps to ensure that the design is optimised and that the capability is adequate. Qualification provides additional confidence through its independent reviews and can bring about improvements to the inspections. Similarly, the open trials provide assurance that the procedure and equipment all work as intended.

Working arrangements. As described in Section 5 above, during the first qualifications, roles, responsibilities and working arrangements were developed. Once these were settled and all those on the qualification side were working together as a **single team**, the problems were addressed more directly and constructively and overcome much more quickly.

Successful working of a single qualification team showed that it was both possible and desirable for the **independent** qualification team to include BNFL Magnox Generation staff. This is allowed by the ENIQ methodology[3] and contributed to a better understanding of the project aims as well as of the inspection equipment and procedures with their strengths and limitations. It also improved communication with the inspection development team and the rest of the project.

Communication. The original intention was to provide for good information flow through documentation. However, this resulted in several documents additional to those required by ENIQ and production of numerous drafts. The preparation and review of these documents slowed down the overall process. Consequently for, later qualification, only the documentation required by ENIQ was produced and information flow was improved through **regular meetings**, which were found to provide the most effective way of ensuring good exchange of information.

Document control. Many documents were circulated in draft form for comment and this created problems with document control. For future qualifications the documentation requirements should be kept as simple as possible and the document issue carefully controlled.

Timescales. The qualifications were all completed without causing delays to the overall repair programme. However, the programme for the qualifications was tight, initially unrealistically so. In particular, it did not allow adequate time for commenting and the subsequent re-issue of documents. Gradually, the estimates of the time required for the qualifications were refined and realistic times included in the overall repair programme.

Ordering of test blocks. Because of the programme tightness and long lead times involved in procurement of test blocks containing realistic defects, in most cases the test blocks were ordered before the relevant TJ was prepared. This meant that the defects finally identified in the TJ as the 'worst case' were not necessarily included in the blocks. In general, this led to additional work needed to demonstrate that the inspection capability is adequate for the 'worst case' defects. The lesson is that the TJ, or at least the part of it which defines the 'worst case' defects, needs to be prepared **and agreed** before the blocks are ordered, ie at a very early stage.

8. CONCLUSIONS

Four inspections were qualified following the ENIQ methodology.

The qualifications were all completed without causing delays to the overall repair programme.

Qualification helped to optimise the inspection design and provided additional confidence in the inspection capability.

After initial problems that could be expected with a new qualification methodology, the ENIQ approach worked well and proved to be flexible.

The structure of the documentation required for qualification should be kept as simple as possible and the document issue controlled carefully.

Five capability statements were produced and endorsed by an external third party.

9. ACKNOWLEDGEMENT

Thanks for the successful outcome of the qualification are due to a number of colleagues in BNFL Magnox Generation, to Chris Pople of CK Solutions, Robin Shipp of Firecrest Consulting, Charles Schneider of TWI, Chris Waites, Brian Hawker and others from AEA Technology and to the members of the Independent Qualification Body, John Whittle, Peter Conroy and Mike Duff.

This paper is published with the permission of the Director Technology and Central Engineering, BNFL Magnox Generation.

10. REFERENCES

1   KJ Bowker, 'Inspection Challenges', Paper 7 of these proceedings.

2   'ASME Boiler and Pressure Vessel Code; Section XI, Rules for Inservice Inspection of Nuclear Power Plant Components', 1995 Edition including 1996 addenda

3   'European methodology for qualification of non-destructive testing (second issue)', ENIQ report No 2, EUR 17299 EN, 1997.

4   'Common position of European regulators on qualification of NDT systems for pre- and in-service inspection of light water reactor components', European Commission Report EUR 16802 EN Revision 1 (1997)

5   LF Exworthy et al, 'An evaluation of the nature and origin of cracking...', Paper 2 of these proceedings.

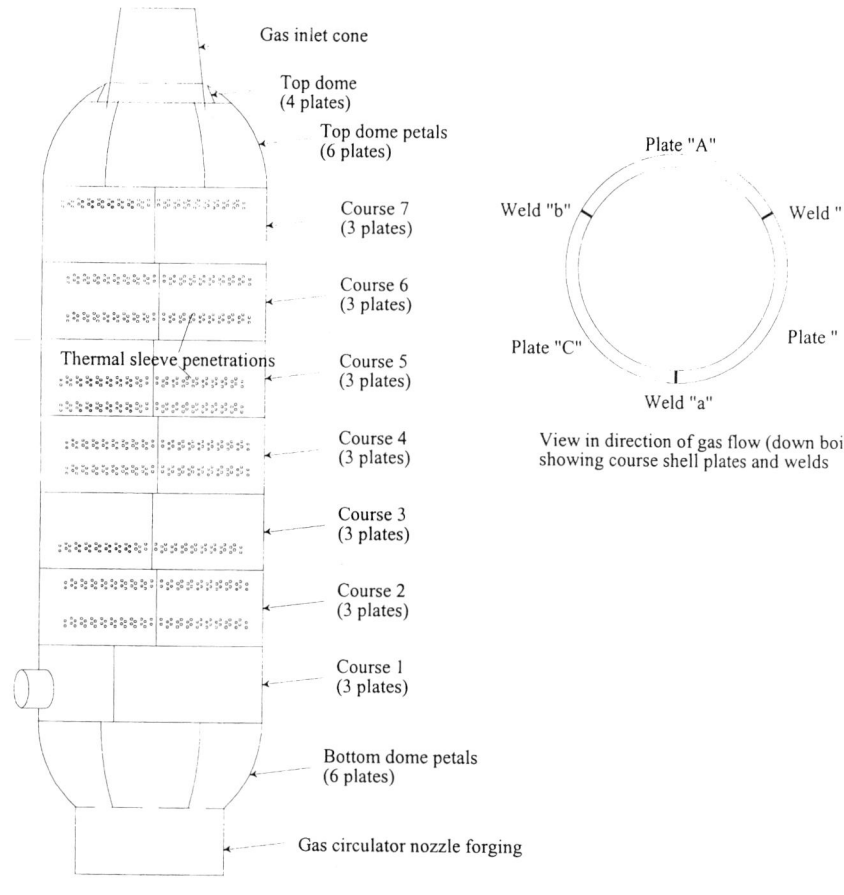

Figure 1: Boiler Shell - schematic

Top of Boiler

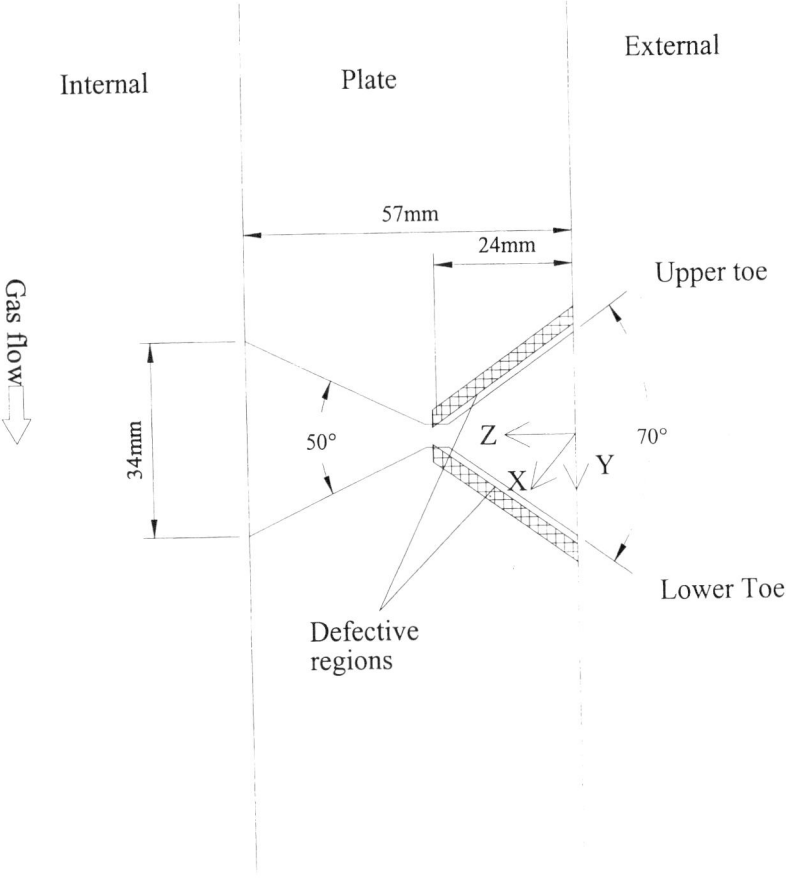

Figure 2: Course 6/7 weld preparation and defect locations

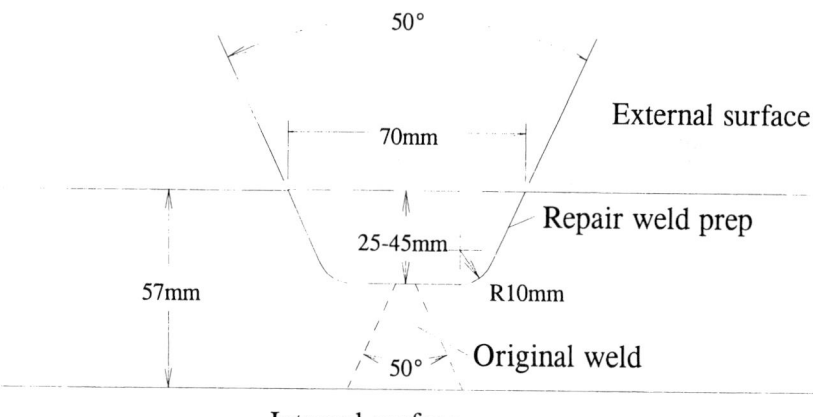

Figure 3: Course 6/7 Repair Weld Preparation

Figure 4: Boiler 2C course 5/6 weld excavation profile (end Elevation)

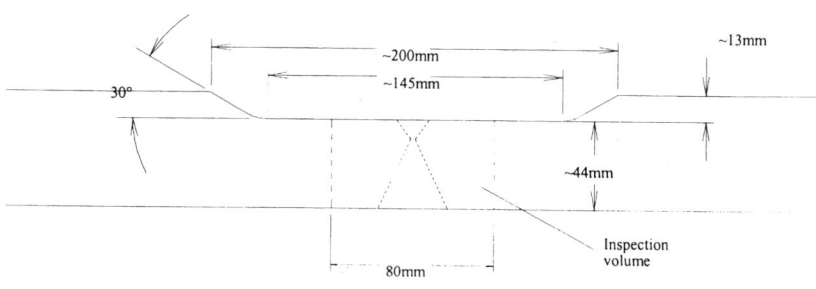

Figure 4: Boiler 2C course 5/6 weld excavation profile (end elevation)

AEA Technology
Energy

VALIDATION
CERTIFICATE

SIZEWELL A REACTOR 2 BOILERS 2A, 2C and 2D REPAIRS TO
WELDS 6/7: VALIDATION OF THE INSPECTION OF THE **EXCAVATION**
LIGAMENTS AFTER MACHINING AND BEFORE WELDING USING THE
AUTOMATED ULTRASONIC NON-DESTRUCTIVE TESTING **PROCEDURE**
FOR THE **DETECTION** OF FLAWS

CERTIFICATE NO: VCP 49   ISSUE: FIRST

This is to certify that the inspection process described in the Procedure identified overleaf has been
applied by operators, whose qualifications conform to the requirements of the Procedure, to Test
Blocks representative of the excavation geometry. These Test Blocks contain flaws which meet the
requirements of TE/SAZ/WI/0037 Issue 2. The results of these tests, which were carried out under
the invigilation of the ME qualification trials team leader and observed by a member of the
Independent Qualification Body representing the IVC, show that the Procedure provides for detection
of flaws as described in TE/SAZ/WI/0037 Issue 2.

| Day/Month/Year |

EXAMINATION PERIOD: 12.04.98 - 13.04.98   Signed (For IVC) ....

Mr C Waites
Head of IVC

EFFECTIVE DATE OF ISSUE: 13.04.98

| FOR NOTES ON VALIDITY SEE OVERLEAF | UNEMBOSSED COPIES ARE NOT VALID |

ISSUED BY
THE INSPECTION VALIDATION CENTRE
ON BEHALF OF
AEA TECHNOLOGY PLC

AEA Technology plc registered office
329 Harwell, Didcot, Oxfordshire
OX11 0RA. Registered in England
and Wales number 3095862

**Figure 5: Qualification Certificate - Inspection of Course 6/7 Weld Preparation
Ligament**

VALIDATION CERTIFICATE

AEA Technology
Energy

SIZEWELL A REACTOR 2 BOILERS 2A, 2C and 2D REPAIRS TO
WELDS 6/7: VALIDATION OF THE NON-DESTRUCTIVE INSPECTION
DATA ANALYST FOR THE EVALUATION OF FLAW GROWTH
MONITORING

CERTIFICATE NO: VC 210   ISSUE: FIRST

NAME & ADDRESS OF EMPLOYER
Mitsui Babcock Energy Limited
Technology Centre
Renfrew
High Street
PA4 8UW
Scotland

This certificate has been awarded to M Venters

Issue is in accordance with AEAT/GWI/13/IVC/11 and the requirements stated in SZA/IQB/1 the Magnox Electric contract
specification document.

Proficiency has been demonstrated as follows:

The above named Data Analyst (DA) has applied the NDT analysis procedure referenced overleaf, to the ultrasonic data
acquired from Test Blocks containing flaw data sets designed to simulate flaw growth. This data constituted a Blind Trial
of the DA's ability to determine flaw growth in the repair weld geometry relevant to welds 6/7 of the Sizewell A reactor 2
boilers 2A, 2C and 2D. In addition, the above named DA has sat two written examinations designed to test his ability to
correctly interpret the procedure and reporting requirements. Based upon the results of the practical trial, the written
examinations and a review of the Qualification Trials Report, it is judged that he is competent to apply the procedure
referenced overleaf.

The IVC has examined the records of the operator concerned and is satisfied that his qualifications conform to the
requirements of the procedure.

Day/Month/Year

EXAMINATION PERIOD: 12.10.98 - 18.11.98   Signed (For IVC) ....
                                                 Mr C Waites
EFFECTIVE DATE OF ISSUE: 18.11.98   Head of IVC

DATE OF EXPIRY: 17.11.99   Signature of Holder ....

FOR NOTES ON VALIDITY SEE OVERLEAF	UNEMBOSSED COPIES ARE NOT VALID

ISSUED BY
THE INSPECTION VALIDATION CENTRE
ON BEHALF OF
AEA TECHNOLOGY

Figure 6: Qualification Certificate - Personnel Qualification for Growth Monitoring

Welder training and qualification for the repair welding of the boiler shells

J TOLAINI
Mitsui Babcock Energy Services Limited, Tipton, UK

ABSTRACT

A team of competent welders was required for the execution of the boiler shell weld repair at Sizewell A Power Station. This paper outlines the selection, training, qualification and prolongation process for 40 welders who were qualified for the manual metal arc (MMA) repair procedure. The monitoring of the weld deposits is also described together with the non-destructive and metallurgical testing acceptance criteria used at each training stage. Details of grain refinement achieved and weld metal volumes deposited are given.

1. INTRODUCTION

1.1 Background
The technique of controlled deposition repair welding was originally developed as a means of controlling parent material heat affected zone (HAZ) microstructure, principally to ensure that a fine-grained HAZ was present after completion of the welding operations to reduce the risk of stress relief cracking in susceptible materials during subsequent post-weld heat treatment.

1.2 Published data
A literature search prior to the development of a specific welding procedure for Sizewell A, showed that the topic had been researched and well documented for materials such as carbon steels(1) and 1CrMo and 2CrMo steels (2,3,4). Indeed, MBESL has been conducting repairs on both as-welded and post-weld heat treated components in the Power Generation Industry using temper bead techniques for over 25 years. Whilst no published work was found for the specific boiler shell alloy steel group, Allen (4) outlined practical parameters for the repair welding of 2CrMo steel plate and these were the parameters used as a starting point for the Sizewell A development trials. However, it did prove necessary to substantially adjust the first layer parameters for the BW87A material.. The development of the weld procedure is described in more detail in Paper 6, these proceedings (5).

1.3 Welder Selection & Training

The welding technique developed for the repair welds on the boiler shell, is a specialised method which is reliant on well defined and carefully controlled weld beads placed in a particular sequence within the excavated area after defect removal .

The welders involved in the welding trials and subsequent weld procedure test pieces were developing the technique and were trained and well practised prior to formal procedure qualification. However they were few in number and therefore the welders to be used at site for the repair needed to undergo specialist training and qualification. The additional welders required were carefully selected for their ability to closely control the MMA process in terms of travel speed and bead placement within a restrictive excavation profile as well as their experience of heavy section welding using the MMA process. A preliminary bead on plate trial was utilised to evaluate the welder's practical ability to maintain a consistent travel speed and bead overlap. Successful welders were selected to progress to the full training programme as detailed in this paper.

It was estimated that the welder training period required would be in the order of 4 weeks (5 day working), with a five day programme of training for the welding monitors.

The training was conducted at the Mitsui Babcock Energy Services Ltd Training facility at Tipton in the West Midlands, under the control of the Welding Services Department using personnel trained in the technique who later attended the Sizewell A site with the welders to then control the actual repair welding activities.

1.4 Welder Qualification

Qualification test pieces were undertaken in BW87A Reproduction material with a simulated repair groove and tested non-destructively in accordance with the requirements of the British and American welder qualification standards, BS EN 287 and ASME IX . Additional metallographic assessment of the parent material HAZ was carried out. The testing requirements of BS EN 287 Part 1:1992 and ASME IX are summarised below :

	BS EN 287-1	**ASME IX**
Either	4 Side bends 120° over 4t	2 Side bends 180° over 4t
or	Radiography to BS EN 1435 Class B or Ultrasonics (for t $\geq$ 8mm) MPI (Optional) to BS EN 1290 Sentenced to BS EN25817 Category B	Radiography to QW191
Plus	Visual to BS EN 25817 Category B & C	Visual to QW 194

BS EN25817:1992	Arc welded joints in steel. Guidance on quality levels for imperfections
BS EN 1290:1998	Non destructive examination of welds. Magnetic particle examination of welds.
BS EN 1435:1997	Non destructive examination of welds. Radiographic examination of welded joints.
BS 3923:1:1986	Methods of ultrasonic examination of welds. Manual examination of fusion welds in ferritic steels.

The welder qualification samples were tested to comply with both codes. The following tests were conducted :

Visual to ASME IX QW 194 and BS EN 25817 Category B & C

Radiography sentenced to ASME IX QW191

Ultrasonics to BS3923 Level 1 (DAC + 14dB) sentenced to BS EN 25817 Category B

MPI to BS EN 1290 sentenced to BS EN 25817 Category B

Parent material HAZ grain refinement evaluation of one macro section

The ultrasonic examination was performed at an enhanced level to reflect the possible on-site inspection parameters.

2. GRAIN REFINEMENT

The welding technique developed is designed to achieve a high degree of grain refinement of the heat affect zone in the parent material.

The technique is essentially a two layer temper bead method, where a low heat input first layer is deposited with 2.5mm diameter electrodes under a controlled deposition sequence and high (~50%) degree of bead overlap. This achieves two objectives, firstly producing a very narrow HAZ, in itself largely grain refined due to the overlapping weld beads, and secondly developing a consistent layer thickness which is important for the penetration of heat from the second controlled layer, sufficient to grain refine the original HAZ. The second layer is deposited in a similar controlled manner as the first, but with 3.2mm dia electrodes and a higher heat input. Heat input ratio between first and second layer is in the order of 1:2.2, i.e. first layer deposited at approximately 420 Jmm^{-1} heat input and second layer deposited at approximately 900 Jmm^{-1} heat input.

A third controlled layer is deposited with the same heat input as layer 2, to provide some measure of tempering of layers 1 & 2, and a fourth controlled layer, again at the same heat input as layer 2, is deposited to provide protection to the parent material heat affected zone from the subsequent welding of the excavation.

There is consequently a requirement to control the deposition of weld beads within established parameters of heat input. Electrode run out ratio is a welding parameter which can be easily measured and is proportionately related to heat input. Run out ratios were therefore used as one of the controls incorporated into the welding inspection and monitoring to ensure compliance with heat input requirements. Electrode run out ratio is defined as the ratio between the length of the weld bead deposited to the length of electrode consumed. It is a control on heat input which is independent of welding current used and is of value in situations where sophisticated measuring equipment may not be available. In this case it was used in addition to heat input measurement and as a practical guide for the welder to achieve certain tangible targets.

Bead overlap is another key factor in achieving the required grain refinement and is defined by the formula below :

% Bead overlap = [a - (b-a)] x 100 / a

where a and b are defined in figure below

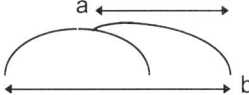

During welding, the monitor measured the bead widths using engineering dividers at one consistent point along the test piece and recorded the readings on the monitoring log sheet.

The welding procedure incorporates a sacrificial run out strip placed at the surface. This enabled the bead placement pattern to remain consistent past the plate surface and out onto the run out strip. This allows a consistent heat flow into the HAZ and grain refinement at and near the surface of the plate is achieved at high levels similar to the rest of the parent material HAZ.

3.   WELD PARAMETER MONITORING

The achievement of the desired grain refinement is dependent on the adherence by the welder to pre-determined welding parameters for the initial two layers of weld onto the parent material.

These are:-

- Current   ⎫ Fundamental parameters.
- Arc Voltage   ⎬ These can be represented by Run-out Ratio
- Travel   ⎭
  speed
- Weld bead overlap

It was therefore essential that for the critical first two layers of weld buttering, the welding parameters were measured and logged. Should an individual weld bead fall outside the set tolerances, then remedial action was taken immediately, by removing the bead by careful grinding. The finished height of the dressed bead was to be level with the previous layer of weld buttering or in the case of a first layer weld bead, level with parent material in the weld preparation.

This required monitoring personnel to be with each welder for the duration of the first two layers, with intermittent monitoring of the next two layers, four in total.   The weld parameters were monitored using proprietary portable arc monitoring equipment, specifically, PAMS V units, see Fig.1. These connected to the welding leads and recorded current, arc voltage, and arc time. The weld length was then measured and inputted into the unit and a printout of average current and voltage values and heat input was then be obtained. These values together with other relevant details were recorded on weld data monitoring sheets (Appendix 1). One of the prime indicators of heat input which can be easily ascertained and checked is run-out ratio and was compared to allowable values in order to accept the run and proceed to the deposition of the next weld bead. The run out ratio was also recorded on the monitoring sheet. Monitoring was conducted from the earliest stage of training to ensure compliance to the target parameters and to enable minor deviations to be corrected early before they became ingrained in the welder's technique.

4.   TRAINING & QUALIFICATION

Training of a team of welders was necessary because the grain refinement technique utilises parameters which would not be commonly used in standard MMA welds. The welders needed to be trained to consistently achieve these parameters for each electrode deposited.

The training strategy adopted was to carefully select welders for assessment based on their previous experience, then practically assess their initial ability to weld positionally to the procedure required onto carbon steel plate material.

Welders who were successful in their assessment progressed to the main training programme. The programme consisted of a detailed induction including the background to the repair and the factors affecting the metallurgical objectives followed by four distinct practical training stages. Each stage had to be successfully completed before the individual welder could progress to the subsequent stage. A summary of the training programme is shown in Table 1. The fundamental technique of electrode deposition were taught and practiced during Stages 1 & 2 and welders gained the ability to repeatably deposit electrodes within the parameters required and also learn the bead deposition sequence and electrode angle for welding the excavation. Stages 1 & 2 permitted considerable practice and repetitive welding to be undertaken allowing the welders to develop the consistency required. Welders progress to Stage 3, the practice plate in alloy steel, was dependent on the confidence of the instructional staff that the welder was able to complete it satisfactorily, based on performance in the previous stages. A flow chart representing welder progress through the programme is reproduced in Table 2.

Welders progressed to the formal welder qualification stage only after their practice welds had been fully volumetrically inspected and evaluated metallurgically.

The nominated Third Party inspection authority for the weld procedure and welder qualifications was Royal & Sun Alliance and the inspectors involvement was not limited to each welders final test piece but he was party to the welders progress through each of the training stages. The inspector also checked the training of the monitors and endorsed their certification.

4.1 Training Stages & Acceptance Criteria
Table 3 outlines the detailed training programme.

4.1.1 Stage 1. Bead on Plate
The welders were be trained to produce welded pads onto carbon steel plate 25mm thick and

long enough to permit the full deposition of an electrode length (300mm). The pads were be deposited in three positions, as shown below :

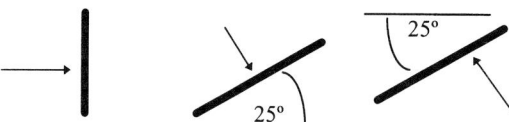

Before a welder was allowed to proceed to the next stage he was required to demonstrate that he was able to deposit visually acceptable pads and maintain consistency of weld deposit within the following parameters in each of the three positions outlined above .

A minimum of 80% of the beads deposited in each layer were required to be within the tolerance of ±5% of the stated run out ratio for that electrode size and layer number. The remainder of the beads deposited in each layer had to fall within the tolerance of -10%,+5% of the stated run out ratio for that electrode size and layer number. This latter, tolerance had been derived from the development work and statistical trials described elsewhere(5), as a working envelope which would produce satisfactory results. Bead overlap was also monitored and was required within the range 40 to 60%.

Training via bead on plate trials continued until the individual welder was able to achieve consistent results within the specified tolerances.

Fig. 2 shows a successfully completed bead on plate test piece, the three, two layer pads being clearly visible.

4.1.2 Stage 2. Carbon Steel Simulated Groove
The next stage in training was the welding of a 25mm thick carbon steel simulated excavation to the training weld procedure S01/0303/27/006 Rev 1 (see Appendix 2). An excavated groove was simulated by preparing backed plate butt welds as shown below :

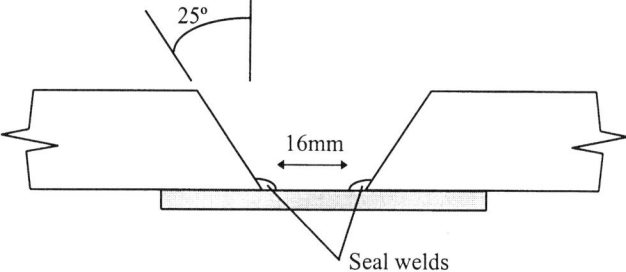

The welders were required to weld the first two buttering layers only, Fig 3 is a completed test plate showing the two layers. The tolerances required on run out ratios were the same as for the bead on plates, section 4.1.1. In addition a macro section was taken to evaluate the bead sequence and establish that a consistency of penetration and run placement was being achieved. Failure to achieve the specified levels resulted in the welding of a further simulated groove for re-evaluation.

4.1.3 Stage 3
Practice Plate in alloy material (Reproduction BW87A or BS1501:271 or equivalents)

On successful completion of the 25mm thick simulated excavation weld, the welders were then asked to complete an alloy plate with an excavated groove 35mm deep as detailed on the weld procedure S01/0303/27/006, and including the sacrificial run off strip along the top edges of the weld preparation.

Weld monitoring was carried out on all beads for the first two layers, and parameters had to achieve the tolerances described in Section 4.1.1 for bead on plate and 40-60% bead overlap. Intermittent monitoring was carried out for layers 3 & 4 and assessed to the same criteria as for layers 1 & 2. On completion of the four buttering layers (see Fig. 4) the welding was interrupted in order to cut a section for metallography, approximately 50mm from the end of the groove. This section was prepared for examination to establish the level of grain refinement achieved. Once the section was taken, the remaining portion of the test piece was returned to the pre-heat temperature and the welder completed the remaining filling passes. The cutting of the test plate after welding the buttering, was carried out to permit metallographic evaluation of the HAZ to be conducted in parallel with the completion of the excavation weld and have the results available at an early stage. The development trials reported elsewhere showed that when welding within the tolerances specified, grain refinement substantially in excess of 90% was achievable and therefore a minimum level of 90% was applied as the requirement from the welder samples. On completion of the full groove, the plate was dressed and ultrasonically tested in accordance with the requirements and acceptance criteria outlined in Section 1.4 of this document. Failure to achieve these criteria would have resulted in a repeat practice test piece after analysis of the reasons for deviation from the criteria laid down.

4.1.4 Stage 4. Qualification Test in Reproduction BW87A Material
The final test piece was the formal welder qualification test in reproduction BW87A material nominally 57mm thick and with a 35mm deep, longitudinal machined excavation. Weld monitoring was conducted to the same extent and acceptance criteria as for the practice test piece as outlined in 1.4 and 4.1.3.

On completion of the weld and surface dressing, the plate weld was surface crack tested by magnetic particle inpection, ultrasonically tested and radiographed to the requirements outlined in Section 1.4. Visual examination was carried out at regular stages during the welding and on completion of the test piece. On completion of the NDT examinations a section was taken and prepared for metallographic examination. Using the same philosophy as outlined in section 4.2.3 above, HAZ grain refinement was measured and reported and a pass criterion of >90% grain refinement was set for the metallurgical assessment. The welding and testing stages of the qualification test pieces were fully witnessed by the Royal & Sun Alliance Engineer Surveyor. Fig. 5 shows a qualification piece during welding and Fig. 6 is a photomacrograph of a typical section taken for grain refinement measurement.

4.2 Training of Monitoring Personnel
A considerable amount of monitoring was carried out during the training and qualification stages by the seven training and supervisory staff who had all been trained in the monitoring techniques used. It was recognised during the planning phase that there would be a number of additional monitors required for the site repair activities. A decision was taken to put all

the welders through a monitoring training course plus an additional six staff, affording flexibility at site for the monitoring activities. All of the personnel trained had to pass practical elements during the training and also sat an exam paper at the conclusion of the course and on successful completion Monitor Competence Certificates were raised.

The monitor training programme is detailed in Table 4.

4.3 Prolongation of Welder Qualification

Conventional prolongation of welder certificates under the rules of BS EN 287 and ASME IX require endorsement of the certificate at 6 monthly intervals but in the case of this specialised technique it was recognised that a period longer than one month without applying the technique was inadvisable without requiring some form of prolongation via a test piece.

It was originally programmed that the welder qualification test would be conducted within a period of one month prior to site welding commencing in order to protect against erosion of the specific techniques in which the welder has been trained. However, the logistics of the testing programme and site delays resulted in most of the welders requiring to undertake a prolongation test. This was a 60mm thick carbon steel test plate with a machined excavation 35mm deep. The welders were monitored during the welding of this test and then the completed weld was dressed and ultrasonically tested. Acceptance criteria adopted during the original qualification piece were applied to this test for the adherence to parameters (monitored data), visual inspection and ultrasonic testing.

No further metallurgical checks were undertaken at this stage.

5.   NON-STANDARD GROOVES

It was recognised that whist the general repair excavation depth on the boiler shells at site was going to be 25mm around the full circumference, there were locations where the excavation had to be machined deeper to remove defective material. Consequently a technique had to be established to ensure that at the changes of excavation shape, the bead placement of the buttering layers was optimised for grain refinement.

Tests were conducted using reproduction BW87A plate with an excavation which was generally 35mm deep with a locally deeper area down to a 50mm depth. Bead deposition sequences were devised and documented for site use, Fig.8 shows a typical deposition pattern. A full scale model was also constructed for reference purposes at site, showing the bead sequence pattern to be used. Metallurgical checks of the test plate HAZ at the changes in depth showed that high levels of grain refinement were still being achieved at > 99%

6.   OUTCOMES

6.1 Qualification plates

Forty welder qualification plates were welded with no failures from either volumetric or visual NDT, nor from the metallurgical evaluation of grain refinement. Fig.7 is a histogram showing the percentage grain refinement obtained from the sample of 40 test plates analysed.

Additionally, from the practice plates (also full grooves) and the prolongation plates there were no failures from volumetric or visual NDT nor from the metallurgical evaluation of grain refinement. One welder was required to repeat his practice plate weld and this was not due to any NDT or grain refinement failure, but the macro section showed poor control of bead penetration into the parent material attributable to incorrect electrode angle.

6.2 Weld Volume

A calculation of weld volume deposited during training on the weld excavations shows that excluding bead on plate training and carbon steel simulated grooves, a total of 241,680 cm^3 of weld metal was deposited equating to 1,902 Kg of deposited weld metal. In contrast to this, the actual volume of weld deposited on the three boiler shells at Sizewell A, was 230,765 cm^3 and 1,854 Kg in weld metal weight. Considering the additional weld metal deposited during the early training stages, there was considerably more weld metal deposited during training than in the actual repair of the boiler shells.

7.   CONCLUSIONS

It was recognised at an early stage in the project that the desired outcome of sound, acceptable weld metal and a fine grained parent material HAZ was reliant on a highly controlled welding technique. The implementation of a carefully planned and designed training programme has shown that the deposition of nearly 2 tonnes of defect free weld metal in the field can be carried out with a high degree of confidence that the metallurgical objectives have also been achieved.

During the welder training programme, preliminary results from the practice plates indicated that the technique was effective and that the welders could achieve the aims set for them. The site phase of the repair therefore commenced with a high level of assurance that the repair would be successfully welded to the stringent quality required.

8.   ACKNOWLEDGMENT

The assistance of many colleagues at Mitsui Babcock Energy Services Limited who were involved with the training programme and the wider group of individuals who assisted in the development of the welding technique is gratefully acknowledged.

9.   REFERENCES

1. Jones, R L : "Development of two layer deposition techniques for the manual metal arc repair welding of thick C-Mn plate without post-weld heat treatment", The Welding Institute Research Report 335.1987, April 1987.

2. Freidman L M : "Repair welding for high temperature equipment in Power Generation and Refinery service", 10th Annual North American Welding Research Conference, October 1994.

3. Freidman L M : "EWI/TWI Controlled deposition repair welding procedure for 1¼Cr½Mo and 2¼Cr1Mo steels", Pressure Vessel Research Council/EWI Workshop, Jan/Feb 1996, San Diego, USA.

4. Allen D, Kelly T : "Cold Weld Repair - Development and Application", 2nd International EPRI Conference, May 1996, Daytona Beach, USA.

5. McDonald E J, Hunter A N R, Bell W : "Specification, development & optimisation of the welding and post-weld heat treatment procedure" Paper 6, Proceedings of the Sizewell A Boiler shell weld repair Seminar, 25 October 1999, Institute of Mechanical Engineers, London

Table 1. Summary of Welder Training Programme

Timescales quoted are typical, actual timescales varied according to each individual's progress rate.

	Activity
Day **1** **2 - 6**	Induction **Stage 1:** Bead on plate trials, including monitoring **Stage 2:** Simulated groove in carbon steel.
Day **7 - 13**	**Stage 3:** Weld a machined groove in BW87A or BS1501:271B Plate. First 4 butter layers only, section end, evaluate metallurgically. Complete fill welding of remaining groove.
Day **14 - 19** **20**	**Stage 4:** Weld a grooved BW87A qualification piece under test conditions. Butter layers plus complete fill. Dress for U/T and section for micro-examination. Metallurgical evaluation.

Table 2. Flow chart of welder progress through the programme

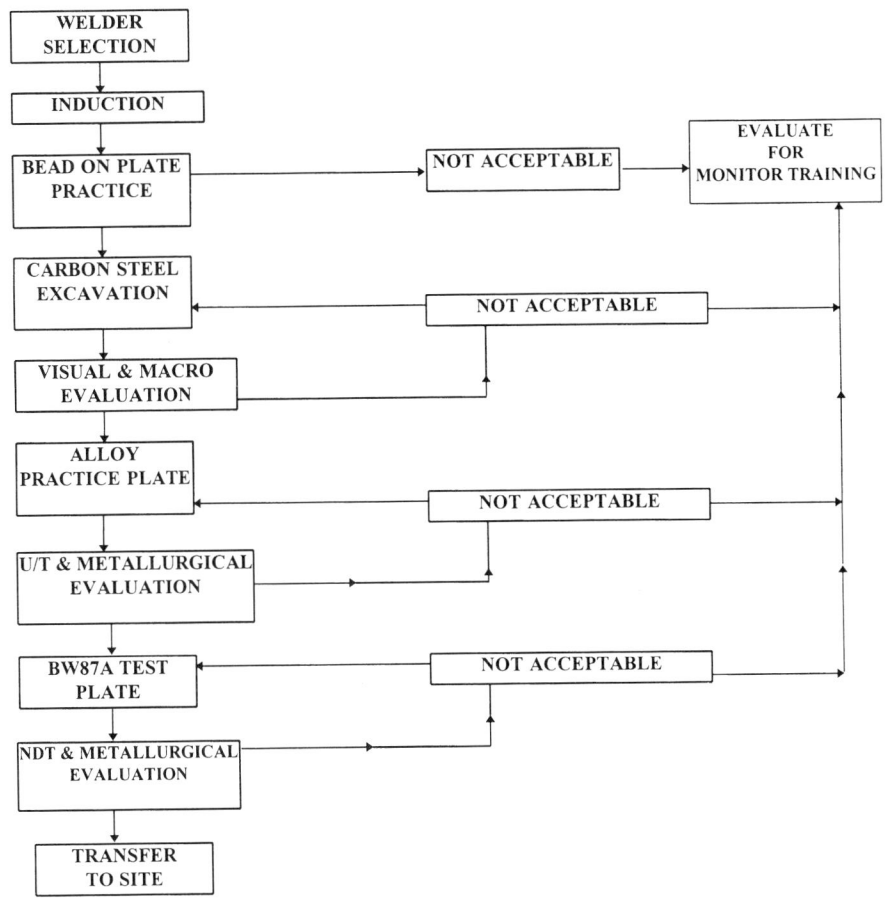

Table 3a. Detail of Welder Training Programme

Day	Activity

1	**Induction**
	
	• Health and safety
	• Background to the repair
	• Requirements for the repair
	    • Metallurgical factors
	    • Welding techniques, bead placement etc.
	    • Grain refinement objectives
	    • Adherence to parameters
	    • Monitoring of parameters
	• Demonstration of PAMS V Unit
	• Site requirements
	
	Setting out in Welding School

2	Practice achieving run-out ratios specified on WPS bead onto carbon steel plate, in three positions, flat, h/v and inclined overhead. 2.5mm dia H1 electrodes: 1:1 ROR, 50% bead overlap, amperage as WPS 3.5mm dia H1 electrodes: 0.7:1 ROR, 50% bead overlap, amperage as WPS.
3	Repeat of day 2 , but measure results using monitoring equipment. Ensure repeatability and consistency of results within permitted tolerances.
4, 5 & 6	Practice in welding two buttering layers on a simulated excavation in carbon steel plate material. Visual examination.

Table continued overleaf

Table 3 (continued). Detail of Welder Training Programme

Day	Activity

7 - 13	**Trials on alloy plate (Rep. BW87A or BS1501:271 or Equiv.)**
	
	Weld four buttering layers on alloy plate with a machined excavation and fitted run-out strips. Instruction on bead placement within the groove. Section of trial plate butter welds ~50mm in from start position. Polish and etch for metallurgical evaluation. Complete welding of remaining groove. Carry out ultrasonic testing after dressing of weld surface.
	
	Plate size: 50mm min thickness
	                  400-750mm long
	                  200-300mm wide
	
	Pre-heat: 200°C gas or electric
	
	Weld Procedure: S01/0303/27/006
	
	Surveillance Monitoring of first 2 layers to be carried out.

14 - 19	**Welder Qualification**
	
	Weld a BW87A Plate with machined excavation. On completion of 4 layers, continue to weld the repair to completion.
	
	Plate size: 50mm min thickness
	                  400-600mm long
	                  200-300mm wide
	
	Pre-heat: 200°C electric
	Weld Procedure: S01/0303/27/006
	Third Party Surveyor witnessing Welding Activities.
	Dress for radiography, ultrasonic and MP Inspection. Carry out NDT inspections.

20	Section test plate weld. Polish and etch for metallurgical evaluation. Raise certification.

Table 4. Detail of Monitor Training Programme

Day	Activity

1	**Induction**
	
	• Health and safety
	• Background to the repair
	• Requirements for the repair
	    • Metallurgical factors
	    • Welding techniques, bead placement etc.
	    • Grain refinement objectives
	    • Adherence to parameters
	    • Monitoring of parameters
	• Demonstration of PAMS V Unit
	• Site requirements
	
	Setting out in Welding School
	
	Practice setting up PAMS V equipment onto the welding cables.
	
	Instruction on completion of monitoring sheets.

2	Instruction on use of measuring equipment, rules, special jigs etc.
3 - 5	Practice monitoring trainee welders and the production of completed monitoring sheets. Instruction on remedial actions (for out of tolerance welds). Written exam paper & issue of competence certificates

Figure 1. PAMS V unit in use monitoring weld parameters

Figure 2. Completed Bead on plate test sample

Figure 3. Carbon steel simulated excavation sample

Figure 4. Alloy steel practice test piece

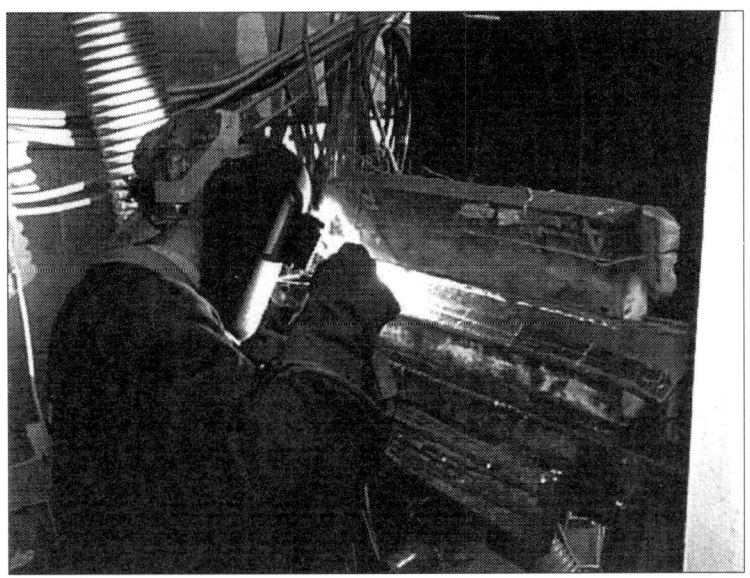

Figure 5. Welder qualification test in progress.
Note the practice strip tacked on below test piece and the electric pre-heat.

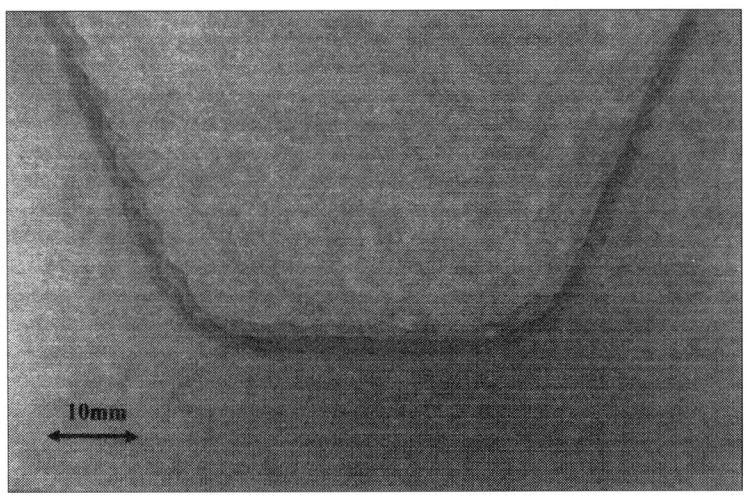

10mm

Figure 6. Photomacrograph of typical qualification test piece

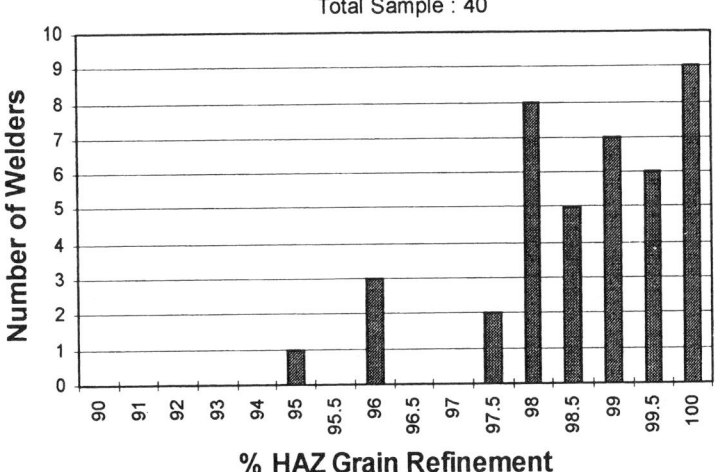

Grain Refinement in Welder Qualification Tests

Total Sample : 40

Figure 7. Percentage grain refinement achieved during welder qualification tests

Sizewell A Repair Weld
Indicative run sequences for areas of groove depth transition

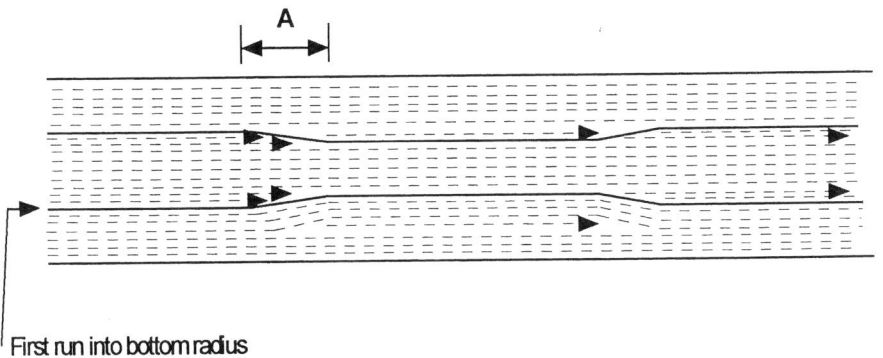

First run into bottom radius

Dimension A : 50mm min for a transition height of 10mm

Figure 8. Run sequence developed for changes in excavation depth

Welder Training Programme, Tipton

Weld Repair Monitoring

TEST PLATE NUMBER :										
Electrode Type:				Batch Number:			2.5mm			
							3.2mm			
Welder No		Welder			Plate Batch No					
Layer No	Run No	Elec Dia/mm	A	V	t (sec)	L (mm)	Bead Width	Width including previous bead	Heat Input J/mm	Stub Length

Monitored By: ....   Date: ....

Appendix 1. Weld Data Monitoring sheet

Mitsui Babcock Energy Services Limited

P.O.Box 8, Birmingham New Road, Tipton, DY4 8YY.
Tel. 0121 530 5000 Fax. 0121 530 5103

Weld Procedure N°	Type & Method of Weld	Welding Positions
S01/0303/27/006 Rev 1 Page 1 of 2	Repair Weld - Welder Training Procedure Manual Metal Arc (111)	Horizontal-vertical

Design/Qualification Code:	Thickness	Outside Diameter
	All	All

Materials	Preheat Temperature & Method	Method of Inspection
BW87A or BS1501:271	200°C minimum preheat 250°C maximum interpass Electric or gas Monitor by T/C or templestik	Visual, MPI, Radiography, Ultrasonics Apply as per contact requirements.

Weld Surface Finish	Post Heat Treatment	Report N°
Dressed to suit N.D.T. requirements.	None required	Not Applicable

Weld Materials

Babcock H1 Electrodes to BWPL Specification BE 0 052

Procedure

1. Dress the welding edges and adjacent areas to clean, bright metal. Ensure that the areas are grease and oil free.
2. Preheat to required temperature. Fit run off plate around top of weld preparation, using light TIG tacks in the weld preparation to S06/0103/26/001 Rev 0.
3. Apply first layer of buttering across the whole face of the excavation using 2.5mm dia electrodes, a 1:1 run out ratio*, and a bead overlap of at least 50%. See sketch overleaf.
4. Carefully visually examine the deposit, lightly dressing spatter and any high spots on the weld.
5. Apply a further layer of buttering across the whole of the first layer, using 3.2mm dia electrodes, a 0.7:1 run out ratio*, and a bead overlap of 50%. This second layer shall extend onto the run off strips.
6. Deposit a further two layers of buttering as per 5. above.
7. Complete welding the repair excavation using 3.2 & 4.0mm dia electrodes taking care not to impinge onto parent material with these filling runs. Filling runs shall be bead runs with minimal weave of ≤2X the core wire diameter. Final bead shall be central in the weld reinforcement.
8. Carry out dressing for final NDT.

Note : Inter-run cleaning is of particular importance and shall be scrupulously applied with suitable tools and heavy duty power rotary wire brushes.

* Run out ratio is defined as the ratio of the length of deposited bead to the length of electrode consumed. eg 0.7:1 equates to 210mm of deposit from 300mm of electrode length (50mm stub).

Typical Preparation

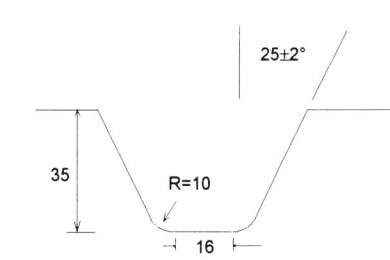

Welding Conditions

Consumable	Babcock H1	
Polarity	DCEP	
Size/mm	Current/A	Electrodes to be stored
2.5	87-100	and dried to ≤5mlH₂/100g
3.2	128-142	as per WIG 15-01-1.
4.0	160-190	

Prepared by:	Checked by:	Date:
Julio Tolaini	Martyn Fletcher	22.02.98

Revision N°	Revised by:	Revision Date:
1	Julio Tolaini	9-4-98

Appendix 2. Test Piece Weld Procedure (Page 1)

⊘ Mitsui Babcock Energy Services Ltd

P.O.Box 8, Birmingham New Road, Tipton, DY4 8YY.
Tel. 0121 530 5000 Fax. 0121 530 5103

Weld Procedure N°
S01/0303/27/006 R1
Page 2 of 2

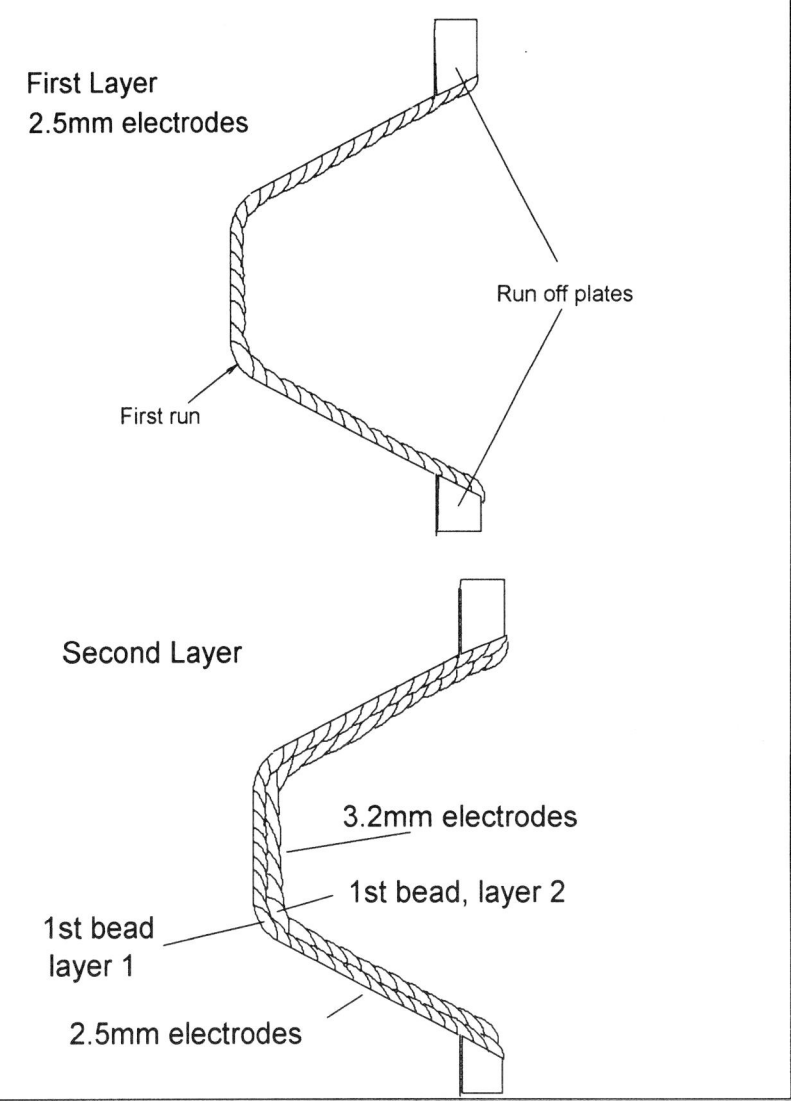

First Layer
2.5mm electrodes

Run off plates

First run

Second Layer

3.2mm electrodes

1st bead, layer 2

1st bead
layer 1

2.5mm electrodes

Appendix 2. Test Piece Weld Procedure (Page 2)

S690/013/99

The role of third party independent support in providing additional assurance on repair quality

E G TAYLOR and **K A SHORT**
BNFL Magnox Generation, Berkeley, UK
A BORLAND
P B Kennedy & Donkin, Manchester, UK

ABSTRACT

The Sizewell A boiler repairs represented a unique challenge to BNFL Magnox Generation in that such extensive repairs, following a prolonged period of service, had not previously been attempted. In support of the Boiler Repair Project, Independent Review and Verification Bodies ("Third Party" organisations) were appointed for the following reasons:-

- This approach was consistent with the requirements of the ASME codes with which the repairs were compliant.

- The Third Parties provided the Regulator with additional assurances on the likelihood of success.

This paper describes which organisations were selected to provide Third Party independent support and their roles and contributions towards achieving a successful outcome to the repairs. The Third Parties have contributed significantly to the Project in the areas of: design appraisal, technical assessment, quality control, code compliance, site surveillance and inspection qualification. As a result of their work, additional confidence has been provided to the Regulator and to Magnox's own internal assessors on the quality of the repairs and the safety of the boilers on their return to service. For the Boiler Repair Project, the Third Party process has been considered to be cost effective in that little rework has been necessary, few delays have been incurred throughout the Project and the boilers have been returned to service immediately following completion of their recommissioning.

1. INTRODUCTION

The Sizewell A Reactor 2 boiler repairs were unique in that such extensive repairs, following a prolonged period of service, had not previously been attempted. The repair strategy for the boiler shells included Post Weld Heat Treatment (PWHT) but no overpressure test on completion. Therefore, the safety of the boilers on their return to service has depended on the implementation of well controlled repair procedures and Non-destructive Testing (NDT) of well established capability. As a result of this distinctive background, Magnox determined that the repairs should be conducted in accordance with the ASME Boiler and Pressure Vessel Codes and that the inspections carried out on the repairs should be "qualified" to ensure the adequacy of their performance. The first of these decisions prompted Magnox to appoint an Independent Third Party Inspection Agency (ITPIA) and an Independent Design Appraisor" (IDA); these roles being equivalent to ASME's "Authorised Nuclear Inspector" and "Registered Professional Engineer" respectively. As a result of the decision to qualify the inspections, an Independent Qualification Body (IQB) was set up to oversee Magnox's qualification process which included testblock trials prior to carrying out site inspections.

It was appreciated at the outset that the boiler repairs represented a significant technical challenge; a repair welding technique had to be developed and implemented which would not crack the underlying original weld metal or parent plate. Magnox therefore decided to appoint external technical specialists to conduct Independent Peer Reviews (IPR) of their proposals. As well as providing an added level of confidence in the prospects of successfully repairing the boilers, the detailed technical assessments also provided an up front investment in that delays in clearances of safety submissions at the back end of the programme could be avoided.

Thus, at an early stage of the programme, a number of Independent Review and Verification Bodies ("Third Party" organisations) were appointed. Their work had the overall objective of demonstrating the adequacy and completeness of all Project activities and essentially breaks down into four main aspects:

- Before commencing welding, independent support has been sought on the adequacy of the technical development work and the likelihood of successful implementation.

- On completion of the repairs, independent judgements were required on the achieved levels of success.

- Throughout the Project, a high degree of surveillance has been applied to ensure adherence to the specified codes and method statements/quality plans.

- A number of important technical issues have been considered in depth to ensure the best solutions are implemented.

In setting up these arrangements, it was recognised that the roles of the Third Parties had to be clearly defined, that they must preserve their independency throughout the Project and that they must work together as a coherent team. This paper describes who the Third Party organisations were and how they worked in their individual roles and together as a team. Finally, the paper reviews the considerable value brought to the Project by this Third Party work, in terms of both technical quality and financial savings.

2.   THE THIRD PARTY ORGANISATIONS

To cover the extensive requirements listed above, BNFL Magnox selected Third Party organisations on the basis of their experience and expertise as follows :-

Kennedy and Donkin Ltd. (K&D) were appointed as the *Independent Third Party Inspection Agency (ITPIA)* whose remit included inspection, monitoring, witnessing and audit. Kennedy & Donkin Quality Engineering are preferred suppliers of Inspection and Quality Assurance Services to Magnox Electric having acquired qualifications under ISO 9001, National Certification Scheme for In-Service Inspection Bodies (now EN 45004) and significant 'hands-on' experience in the Nuclear Generation Industry. They commenced their role with an extensive series of audits on the various organisations within the Project and, from an early stage of the work, they were established on site providing 24 hour surveillance.

Royal and Sun Alliance Engineering Ltd. (RSA) have provided *Independent Design Appraisal (IDA)* which comprised a broad based, independent appraisal of Magnox's design of the repairs, determining the degree of code compliance and assessing any technical justifications for non-compliance. RSA Engineering (formerly National Vulcan) were first approached by Magnox in early 1996 to undertake a construction history review of the boiler shells in response to the initial discovery of the weld defects. They were therefore already familiar with the Project background and this, together with their extensive experience of technical underwriting, made them the ideal choice as the IDA. RSA have also fulfilled other Third Party roles: they were asked by the NII to take a particular interest in all NDT proposals, paying particular attention to those inspections not subject to independent Qualification (see below). They have also acted as an *Independent Welding Approval (IWA)* body; ensuring that welding procedures and personnel are approved using the Safety Assessment Federation (SAFed) guidelines. Finally, RSA have been the *Pressurised Systems Regulator (PSR),* in accordance with the UK Pressurised Systems Regulations.

Professor Michael Burdekin of the University of Manchester Institute of Science and Technology (UMIST) and Dr Peter Hart of The Welding Institute (TWI) were contracted to carry out *Independent Peer Review (IPR)* of specialist technical aspects of the repairs. Each is an accredited expert in the fields of metallurgy and welding respectively. Whereas IDA provides the breadth of the appraisal, IPR has involved in-depth assessments, in particular of the welding methodology, PWHT design and implementation, structural integrity and material properties.

In the absence of an overpressure test, the NDT forms a significant part of the safety case for return to service. The inspection strategy for the repaired and profiled welds has therefore included "Qualification" by BNFL Magnox of the inspection procedures, equipment and personnel. The Independent Validation Centre (IVC) of AEA Technology (AEAT) provided an *Independent Qualification Body (IQB)* to monitor and review this qualification process. AEA-T were ideally placed to carry out this role as they were responsible for the "Sizewell B Validations"; by far the most extensive NDT qualifications carried out to date in the UK. The programme has also called for inspection of a number of boiler components which may be affected by the PWHT. Where these inspections have been considered to be of less significance than those of the repaired welds, but still important to the safety case claims, "Capability Statements" have been produced, drawing on existing theoretical and experimental evidence. The Capability Statements have been reviewed and endorsed by NDT specialists from Rolls Royce Marine Power (RRMP).

3. THIRD PARTY WORKING METHODS

The roles of each Third Party Organisation were specified in detail within the contracts placed by Magnox. However, it was agreed from the outset that there would be no technical/quality constraints placed; the Third Parties were free to challenge and check the work of Magnox and their contractors as they considered appropriate. The large number of submissions which required peer review meant that a prescriptive approach was necessary in order to highlight the issues as Magnox saw them, but, at the same time, the reviewers were to be completely free to extend the scope of the review. A formal, documented approach was adopted by Magnox and supported by the NII. This involved a grading system which specified the depth of peer review required from each Third Party on each document and highlighted any specific technical issues which they were to address. This approach worked well and has led to an auditable record of the IPR process.

The NII were particularly keen that the Third Party Organisations should regularly exchange information. Meetings were held at key project milestones, in particular as the various "hold points" approached (see Section 4 below). The NII attended these meetings and offered their own challenges to "the team". At each NII hold point, all Third Party organisations produced a report, defining their contributions and their position with respect to support for release of the hold point. At the same stages of the Project, Magnox produced its own overview reports of the Third Party work which were reviewed by the NII. In addition to these "milestone" reports, the ITPIA and IDA provided monthly progress reports detailing their activities and findings and highlighting any problem areas for resolution by the Project team. These excellent communications provided confidence that the Third Party process was working well, that any overlap was agreed and that there were no gaps.

Having explained how the Third Party organisations were set up and how they worked as a team, the sections below describe, in more detail, the roles and contributions of each organisation and how the Third Party process added significant value in terms of safety, quality and cost effectiveness.

4. COMPLIANCE WITH ASME

All safety submissions included statements on how the proposed work would comply with the ASME Codes. The IDA evaluated the proposals for code compliance and agreed any non-compliances.

It was a fundamental requirement of the Project Quality Programme that all the major participants in the Repair Project be the subject of a formal Quality Audit to ensure compliance with the relevant requirements of ISO 9001, ISO 9002, ASME III, ASME XI and BS 5882. All organisations were, therefore, audited by the ITPIA to an agreed programme. A total of fifty-five non-compliances were raised and consequent corrective actions agreed and closed out as appropriate. In order to demonstrate continuing compliance with the ASME Codes, Quality Assurance activities were maintained during all the Quality Control work to ensure that the requisite standards were adhered to and that the approved procedures were maintained.

5.   THE SAFETY FACTOR

Magnox already has a well established in-house process for carrying out nuclear safety assessments. The Third Party process is distinct and separate from the INSA; it does not provide a substitute but, rather, complements in-house safety assessment. All Safety submissions were made available to Third Parties, in parallel with Magnox's own assessors, in order that they were provided with the widest range of information on which to base their own reporting. The Third Party organisations contributed significantly to the development and implementation of the safety case by: reviewing and endorsing the supporting technical arguments, monitoring implementation of the associated procedures, confirming compliance with codes and recommending where change or further justification was required.

An important feature of the self regulation of nuclear installations in the UK, is the NII's role in providing their consent for the Operator to make changes to the state of the plant and their approval of the associated safety submissions. In the case of the Sizewell A boiler repairs, the NII defined a series of "Hold Points" which could only be released on the basis of providing adequate information to persuade them that the next significant stage of the programme could be safely implemented. Third Party support for the release of these hold points was crucial to making progress with the repairs without introducing delays. With respect to the overall safety case, some ~100 safety submissions have been prepared and assessed. The NII have reviewed the safety arguments following INSA by Magnox's Health, Safety and Environment Department (HSED). However, in some cases the safety arguments are underpinned by technical evidence of a highly specialist nature and even a selective, detailed technical assessment of these by the NII could have taken months. The IPR process has provided the confidence that all judgements and proposals have been technically sound, minimising the requirements on the NII to employ their own external technical specialists. The independent reviews have provided convincing evidence that the boilers were in a sound condition for return to service and NII approval was obtained only three days after the work was complete.

6.   THE QUALITY FACTOR

Quality Assurance is also traditionally an in-house activity; Magnox, its main supplier Mitsui Babcock Engineering Systems LTD. (MBESL) and the main subcontractors (Didcot Heat Treatment and Reekies Engineering) are all "quality companies", well versed in the ethos of QA. The introduction of the ITPIA has represented a substantial additional measure. They acted in an equivalent role to the "ASME Authorised Nuclear Inspector" and their responsibilities included the following:

- Reviewing and commenting upon Project Quality Programmes raised by Magnox and the main contractor Mitsui Babcock Energy Services Ltd. (MBESL)

- Reviewing and commenting upon contractor and sub-contractor Quality Plans raised by MBESL and their sub-contractors.

- Witnessing or verifying site implementation of approved Method Statements in accordance with Magnox's Work Control Cards.

- Carrying out Quality Surveillance activities both on and off site to verify compliance with approved procedures, Quality Plans and Work Control Cards.

- Documenting all non-conformances observed during surveillance and reporting all such occurrences to Magnox Project Quality Engineer. Ensuring that all such non-conformances were satisfactorily 'closed out'.

7.   SITE SURVEILLANCE

The role of the ITPIA on site developed with the on-going repair work. During the whole period of the repair the ITPIA reviewed all the Method Statements issued and checked them for compliance with the relevant Stage Submissions. All relevant Quality Plans issued by MBESL were subject to review and mark up by the ITPIA. No restrictions were placed upon the ITPIA on the content or areas covered by Quality Plan mark up.

As the activities defined in the Quality Plans commenced, the ITPIA were involved with all witness points marked up on the Quality Plans. All activities witnessed were checked for relevant Method Statements, operator qualifications, Work Control Cards, equipment calibration certification and consumable controls. All non-conformances observed during these witness points were the subject of the issue of a formal Non-Conformance Report for action by Magnox and MBESL. Where necessary, work was suspended until a satisfactory course of action was agreed. These Non-Conformance Reports were maintained by Magnox, subject to a formal 'close out' procedure and endorsed by the ITPIA upon satisfactory resolution of the problem.

In addition to the formal survey of the activities delineated in the Quality Plans, the ITPIA also carried out random patrol surveillance activities on all actions being carried out on site. During these activities spot checks were made on on-going works to check for documentation control, equipment and operator qualifications in addition to highlighting any unsafe or inadequate practices observed. The manning levels used by the ITPIA reflected the levels of work being carried out by MBESL and their sub-contractors. As the work content increased, the ITPIA manning levels reflected this by providing seven day, twenty-four hour coverage during all the preparatory works, welding operations and subsequent re-instatement activities. The areas of inspections included external and boiler internal examinations, inlet ducts, pre and post weld heat treatment equipment, machining processes, welding processes, instrumentation and control equipment, strain gauging and temperature monitoring equipment, non-destructive testing, dimensions controls, insulation controls, supplies and consumables. The inspections and examinations began prior to the commencement of preparatory works through to final non-destructive testing after the trial period following return to full generating capacity of all three boilers.

8. THE DESIGN CHALLENGE

The IDA provided a broad based independent appraisal of Magnox's design of the repairs ensuring overall adequacy of design and compliance with the ASME code. In ASME terms, RSA have acted as the "Registered Professional Engineer", certifying "Design Reports" as appropriate.

The first step in designing the repair was to design a weld repair profile which was acceptable to the IDA. This was readily achieved but the IDA insisted on an increased programme of sample replication to ensure that there would be no welding over significantly cavitated material. This required some local widening and deepening of the groove geometry to remove all defective material which, in turn, led to implications for the welding implementation and the NDT. This design feature provides a good example of the interaction required between all Third Party organisations in achieving a successful outcome. In a similar manner, the IDA confirmed their acceptance of the blended profile in weld 5/6 of Boiler 2C from the aspects of thermal and pressure stresses and inspectability.

The design of the welding and PWHT was also subject to IDA review but the main issues were of a detailed technical nature and therefore primarily involved UMIST and TWI (see below). However, implementation on site required the IDA (and the Peer Reviewers) to be satisfied with the arrangements for storage and control of welding consumables to ensure low hydrogen levels. Furthermore, in order to reduce the risk of plant damage during PWHT, major temporary plant modifications were implemented to reduce loads on components and avoid excessive temperature gradients and strains during thermal expansion. The welded brackets which normally support the superheater tubes and their support beams were deloaded by providing alternative support for the weight by suspending the tube bank from the boiler inlet cone; this also ensured that the tube bank would not be damaged due to differential thermal expansion between the shell and the tubes. In order to eliminate any possibility of damage to the boiler inlet nozzle, the outlet gas duct or the reactor pressure vessel, the outlet duct was separated above the main gas valve with the assistance of a specially designed duct restraint system. The IDA performed calculational checks on loadings, flexibility and finite element analyses and, based on their extensive experience in this type of process, confirmed their satisfaction with the proposals in advance of commencing works.

Commencement of welding constituted one of the "Hold Points" set by the NII. In order to release this hold point, the IDA had to provide their written endorsement, in particular, declaring their acceptance of structural integrity aspects. (Likewise, the ITPIA had to confirm "plant readiness" in writing). Following completion of welding, PWHT and inspections, further hold points were encountered at the stages of initial pressurisation (no nuclear heating) and full pressurisation and power operation. Key activities involving the IDA during these stages of return to service included: the consideration of strain monitoring data and confirmation of acceptable behaviour of the load changes and hysteresis values in the duct hangers.

The vital role of inspection in supporting the safety case for return to service has already been mentioned. An important aspect of the IDA work was to provide representation (together with K&D, UMIST and TWI) on the Project Defect Assessment Panel (PDAP) which considered all inspection findings, as they became available throughout the repair programme, and sentenced any "flaws" reported.

9. THE TECHNICAL CHALLENGE

Magnox, MBESL and the main subcontractors have provided technical specialists to the Project, many of them dedicated full time to achieving a successful outcome. However, the Peer Reviewers (primarily UMIST and TWI) have been able to stand back, with no programme pressures of their own and use their experience to assess the approach, challenge the detail, request evidence and seek justifications for what Magnox were proposing to do at each stage. The key fundamental issues have been addressed in other papers and the vital inputs from the IPR process are as follows:-

One objective of the repair weld procedure has been to minimise the risk of reheat cracking in the newly formed base metal heat affected zone. This has been achieved by the use of a controlled deposition welding technique (to produce a refined HAZ while minimising residual stresses) and an optimised PWHT procedure. Both UMIST and TWI were closely involved in the development of the welding technique. A minimum amount of coarse grained material has been achieved by using low heat inputs and by strict control of welding parameters which have been reviewed in detail and ultimately supported by the Peer Reviewers. Hydrogen dispersion has been achieved by increasing preheat from 150 °C to 200 °C; a significant procedural change resulting from the IPR process. The electrode type was also changed following detailed debate on the balance between strength margins and the risk of reheat cracking. During the welding of Boiler 2D there was a local preheat excursion due to the failure of one heater. This had to be justified and accepted by both UMIST and TWI.

Care had to be taken with the design of the PWHT procedure as, investigations by TWI confirmed that the likely source of original cracking of the site welds was due to a PWHT procedure which was not optimised in terms of creep ductility. A higher soak temperature and a broader heated bandwidth were agreed and strict control of heating and cooling rates were recommended, in particular to avoid spending too much time in the "ductility trough". In order to achieve (and demonstrate) acceptable temperature profiles and minimise thermal gradients in practice, extensive insulation was installed and temperature monitoring was carried out by ~1000 thermocouples.

In order to convince the Peer Reviewers that the boilers would be re-entering service with acceptable material properties, it has been necessary to conduct a substantial programme of materials testing. Pessimistic values (with respect to R51 recommendations) for lower bound fracture toughness (upper shelf) have been agreed, based on laboratory test results. The Peer Reviewers have had full access to the materials testing programme results and have confirmed their satisfaction with the way this work is proceeding.

The remaining key technical challenge to address was that of developing and implementing a programme of inspections which would have a very high probability of detecting defects of concern. This is now briefly reviewed in Section 10 below.

10. INSPECTION QUALIFICATION

As the NDT forms a significant part of the safety case for return to service, it was decided that Magnox would demonstrate the performance of the most important ultrasonic inspections using experimental testblock trials and theoretical evidence. Previously, such a formal approach has only been applied to Sizewell B where ultrasonic inspections of the main components of the primary pressure circuit were validated at AEAT's Inspection Validation Centre (IVC) at Risley. (It should be noted that "Validation" is an ASME term which is equivalent to the European "Qualification".) Since the Sizewell B validations, general guidelines on inspection qualification have been developed in Europe by the European Network for Inspection Methodology (ENIQ).

The extent of qualification recommended by ENIQ ranges from no qualification, where an inspection is routine and straightforward, to the provision of a full Technical Justification (TJ), based on experimental trials and supporting theoretical evidence, for those inspections with a significant degree of difficulty, novelty, commercial importance or safety implications. In accordance with this philosophy, the qualification of the Sizewell A inspections varied according to the perceived significance; details are presented elsewhere in these proceedings. The most extensive qualification was applied to those automated ultrasonic inspections which were carried out immediately following the completion of the weld repair and repeated immediately after PWHT. These inspections, carried out while the shell metal was still at ~150 ° C, included a technique known as Time-of-Flight Diffraction (TOFD) which utilised pairs of angled beam probes and was used to monitor for any defect growth which may have taken place during PWHT. Qualification in this case involved "blind" trials where the operators had no advance knowledge of the artificial defects which had been implanted into the test block. The blind trials confirmed the performance of the techniques and equipment already established by "open" trials and TJ, and were also used to formally qualify the individual operators. Both the open and blind trials were witnessed by the IQB and, following their vigilance during monitoring of the interpretation of data, it was considered necessary to supplement the blind trials with a written examination for the operators.

The IQB, with their special expertise and experience in NDT, reviewed and helped to shape the TJs, while preserving their independence throughout. Also, together with the IDA, they were asked to independently review those safety submissions associated with the inspection programme. (This was in addition to INSA). In this way, the IQB and the IDA were also employed as part of the IPR process.

For completeness, it should also be noted that inspections carried out to demonstrate that the PWHT did not adversely affect adjoining boiler plant structures were additionally subject to independent review. A total of five Capability Statements were prepared to provide detailed justification of the claimed performance of the NDT, based on a wide range of modelling results, existing test trial data, and experience gained from similar inspections in the past. These Statements were scrutinised by Rolls Royce Marine Power (RRMP) who endorsed the claims made on both the ability to detect and size defects of significance.

11.   THIRD PARTY SUPPORT FOR PLANT REINSTATEMENT

It has always been recognised that a substantial effort would be required to reinstate and recommission the plant. During this period of re-instatement, control of the work moved from the Boiler Repair Project to the Station's Reactor 2 Recommissioning Committee with technical support provided by Magnox's Technology and Central Engineering Division (TCED).

Following completion of the repairs, the plant was re-assembled; essentially a reversal of the processes of deloading the boiler tube banks and separating the gas ducts. The IDA continued to perform an essential role in reviewing and supporting the method statements for putting the plant back together in this way without causing any inadvertent damage. The return to service was effectively achieved in two stages (2 hold points); the boilers were pressurised to 150 psi (1.03 MPa) and the temperature was raised to 160 ° C(no nuclear heating) and finally the plant was returned to normal full power operation. During both of these stages, extensive plant monitoring was carried out in the form of strain gauging, CO_2 leak detection and duct hanger monitoring (supplemented by acoustic monitoring, vibration monitoring and normal boiler instrumentation checks). The ITPIA continued their site surveillance activities throughout these final stages of the programme, approving detailed arrangements for implementing plant monitoring and witnessing installation of equipment and data collection.

In parallel with these site activities, the peer review process continued throughout the plant reinstatement period, in particular with the assessment of results from the ongoing materials testing programme and the resolving of outstanding minor issues on the demonstrated capability of the NDT. Following a short period of full power operation, the plant was shut down in order to carry out planned re-inspections. This completed the extensive inspection programme and the PDAP confirmed the acceptability of the inspection results. These "final" inspections, once again, provided assurance that the repair has been successful.

12.   SUMMARY

The involvement of the Third Party Organisations can be seen as providing a substantial contribution to the successful outcome of the repairs. They provided an additional tier of review and control to ensure the project selected the best repair route, and it was implemented correctly.("Do the right thing, and do it right")

Whilst the cost of this additional tier can be seen as substantial, it should be seen as an investment. The returns are a reduced level of rework(practically zero in this instance),and increased confidence from the assessors and regulators(hence a much reduced level of delays when clearances and approvals were required)

These returns were only achieved by careful selection of the Third Party Organisations, to ensure they had the necessary competencies and expertise, and by careful integration of their involvement within the project.

13.   ACKNOWLEDGEMENT

This paper is published with the permission of the Director Technology and Central Engineering, BNFL Magnox Generation.

S690/014/99

Conclusions of the project and lessons for the future

C J MARCHESE
BNFL Magnox Generation, Leiston, UK

ABSTRACT

The successful repair and return to operational service of the three boilers at Sizewell A enabled Reactor 2 to start generating electricity again on 27 March 1999 after a 39 month shutdown. Reinspection of the repaired areas in May showed no change after commissioning the boilers and following a period of operation at full output.

The success of the project has been reviewed to draw out the lessons to be learnt for future work of this type. Key elements were the integration of Mitsui-Babcocks technological and resourcing capability into BNFL Magnox Generation's project, the technical capability of the company, the support from Third Party and Peer Reviews and the role of the HSE. What was achieved was the return to service of a commercially viable reactor system to the desired quality and carried out to World Class Safety standards.

1. INTRODUCTION

On 28 December 1995 Sizewell A R2 was shutdown to complete a 30 year PSR programme of work to gain permission to operate for a further ten years. During inspections of the boiler shells, four significant weld defects were found in three of the four R2 boilers. In September 1997 the Magnox Electric Board approved a scheme to repair these defects to modern standards and complete the PSR programme to return this reactor to service.

2. PROJECT DEVELOPMENT

Considerable effort was taken to set up the project structure with a Project Management Board integrating the technological and resourcing capability of Mitsui Babcock into BNFL Magnox Generation and several audits have shown that this was achieved successfully.

The working arrangements used were built around groups responsible for the technical, licensing, site implementation and quality assurance reporting to an overall Project Manager. The Project Manager was responsible to the Station Manager as owner of the scheme within Magnox Electric plc and then BNFL Magnox Generation Business Group. Free access was given to the quality group to take part in the other three groups which ensured quality was not added after stages of the project had been completed but integral to each stage. The separate group did, however, provide a means to plan quality assurance, inspection verification and validation issues in advance. This was a considerable success of the project and the low level of rework was due to this approach.

The Project Management Board had involvement of Magnox Electric Directors from its inception to review with the Station Manager and the Project Manager the status of the project. Again this proved useful and successful as the project changed in nature with the developing technical challenges.

The licensing group has to manage the company arrangements under the nuclear site licence and manage the interface with the NII. We have considerable experience with the nuclear safety management of such projects. A choice was made to have an extensive stage submission approach to seeking regulator consent to the progress of the work. Good interfaces with the NII were achieved that will serve as future models and little if any delay has been caused by the Regulatory interface during the project. The structure of the project was built around quality and safety which has been covered in previous papers.

3. TECHNICAL CHALLENGE

The project faced a number of significant technical challenges driven by a combination of circumstances such as the shell materials, their age, the need for in-situ working and the demonstration of successful repairs.

Of particular note were:

1. The development of a welding technique that reduced the stress induced on the aged material and demonstrated that large repairs could be made on such materials.

2. The integration of a range of validated NDT techniques both cold and hot, the latter which proved highly beneficial to the reinspection.

3. The optimisation of the heat treatments both as pre-heat and further PWHT. This was a substantial undertaking and demonstrated that modern heat treatment and design techniques can be deployed in situ.

While the early project internal target in December 1997 was to complete the repairs by November 1998 a major technical decision had to be made around April 1998, following quantitative inspection to size the defects and diagnosis of the origins of cracking combined with further materials sampling information. Although these confirmed the original identified cause, cavitation remote from original defects challenged the success of being able to heat treat these areas. This risk had to be eliminated by removing all the outer cavitated weld material and heat affected zone by machining a groove into each weld completely circumferentially – 20 metres instead of between 1 to 4 metres. At the same time the remaining material below the groove exhibited a microstructure which had to be treated to a higher temperature than originally thought needed, an increase from 650°C to 675°C.

These two decisions were needed to guarantee technical success and increased the internal and external amount of work on the boilers considerably. This took time to assess and was reviewed and agreed by the Magnox Electric Executive in October 1998 prior to work commencing at site.

4. PROJECT OUTCOME

The repairs were complete in February 1999 and the boiler recommissioning carried out. No remedial work was necessary as the repaired welds and profiled defect met ASME standards or better with no residual defectiveness. The two month period after completion of the repairs was taken up by a particularly intensive part of the project. Following acceptance of the NDT analysis of the repairs, reinstatement of the plant was completed. Some of this had started at the end of 1998 when the first boiler repair had been finished but reached peak activity in February. Boiler tube hydraulic testing, rewelding of external feedwater and steam connections, boiler relagging, reinstatement of internal boiler tube supports, boiler drums and gas duct connections to the reactor all had to be completed and tested.

Leak testing of the pressure circuit first in air then in CO_2 at increasing pressure stages was accompanied by monitoring of all the additional instrumentation, strain gauges and leak detection apparatus, specially designed for these boilers. Continuously throughout this process all the project physical infrastructure, scaffolding, insulation, heating system, lifting systems had to be removed from the reactor area to permit a safe start up. After successful recommissioning we received Regulator consent to pressurise the boilers and reactor on 4 March. We received consent to start up Unit 2 on 24 March, were critical on the 25 March and synchronised to the grid on the 27 March 1999.

5. THE SAFETY RECORD

A key success was the safety record. In over 1 million man hours of work involving a core team of 300 contract staff and over 1000 individuals beside station staff, we achieved a zero LWCR (Lost Work Day Case Rate i.e. no lost time accidents), no statutory reportable events and only one minor fire remote from the repair work while cutting up waste boiler components.

This was despite the hazards due to:

    1 6000 boiler entries in air hoods

    2 8 boiler asbestos strips to a total of 100 tonnes

    3 high heating temperatures

    4 1000 tons of scaffolding

The project has individually gained a ROSPA Gold Award this year and this must in part be due to the use of modern quality management systems which organised the project from day one with the aim of achieving safety for all as well as the required quality of the total work. The Construction Design and Management Regulations (CDM) were actively used to the benefit of the project to achieve the right level of safety control. During this time Sizewell A Power Station achieved Level 8 on the International Safety Rating System (ISRS) and the implementation of systems to this Standard made a contribution to safety improvements.

6.   REINSPECTION

The return to service power raising phase was completed successfully and full power on Reactor 2 was achieved early in April 1999. All monitoring systems showed expected behaviour of the repaired welds and the boiler commissioning period continued at full power until May when a scheduled shutdown took place.

With the plant shutdown the commitment to reinspection of the repair was undertaken with the reactor under hot pressurised shutdown conditions. These NDT inspections showed no change in the condition of welds and the Unit was returned to service in five days operating at full power from this time.

7.   LESSONS FOR THE FUTURE

These can be summarised as follows:

1. The project arrangements where a dedicated Project Manager, Site Manager and Technical Manager reporting through to the Station Manager, accountable to a Project Management Board of Senior Managers is a model that worked well.

2. The project arrangements for both nuclear and conventional safety and the relationship with the Regulator achieved an outstanding performance.

3. The quality arrangements for the repair based on Independent Third Party inspection agreed with the Regulator proved a valuable contribution to the acceptance of a safety critical task and eliminated the commercial risk of not achieving the required quality.

4. The use of the Construction Design and Management Regulations combined with systems audited by Det Norske Veritas (for the International Safety Rating System) achieved a very high safety record.

5. The combined technical capability of BNFL Magnox Generation Technology and Central Engineering Department and Mitsui Babcock working in a partnership mode demonstrated that extensive repair work on aged materials of a pressure circuit can be carried out successfully.

Considerable benefit was gained by repairing these boilers and Unit 2 has performed well on completion of the project.

8. ACKNOWLEDGEMENTS

The successful outcome of these boiler repairs was brought about by the co-operation of many organisations. The support of the following is acknowledged as contributing to this successful outcome:

Mitsui Babcock Energy Services Ltd.
Mitsui Babcock Technical Centre, Renfrew
Reekie Machines Ltd.
Didcot Heat Treatment Ltd.
Finchams Insulations Ltd.
Dixons Scaffolding Ltd.
Deborah Services Ltd.
Kingtime Electrical Services
FB Taylor (Cable Contractors) Ltd.
Balfour Kilpatrick Ltd.
JT Pegg and Son Ltd.
South East NDT Ltd.
Kennedy and Donkin Power Ltd.
Royal Sun Alliance Engineering
Professor F.M. Burdekin FREng FRS (University of Manchester – Institute of Science and Technology)
TWI (The Welding Institute)
The Institute of Mechanical Engineers
BNFL Magnox Generation staff at Berkeley Centre and the staff of Sizewell A Power Station.

This paper is published with the permission of the Director Technology and Central Engineering, BNFL Magnox Generation.

Late Submissions

Specification, development, and optimization of the welding and post-weld heat treatment procedures

A N R HUNTER and **W M BELL**
Mitsui Babcock Energy Limited, Glasgow, UK
E J McDONALD
BNFL Magnox Generation, Berkeley, UK

ABSTRACT

A manual metal arc procedure was successfully developed and qualified for the Sizewell 'A' boiler repairs. The basis for this procedure was a 'temper' bead technique which allowed the creation of a fine grained HAZ in the repair weldment.

A local post weld heat treatment procedure was developed in which thermal gradients were minimised and a ramp rate, close to the maximum allowable code values, was achieved. This, in combination with the fine grained repair weld HAZ, gave a high level of confidence that the boilers could be repaired successfully and without the reoccurrence of stress relief cracking.

1.   INTRODUCTION

As discussed in Reference 1, the defects found in Sizewell A boilers, Fig. 1, were identified as stress relief cracks associated with the 6/7 circumferential weld seams in units 2A, 2C and 2D and also the 5/6 seam in unit 2C. This cracking occurred primarily in the grain coarsened heat affected zones (GCHAZ) of these welds. It was considered that this microstructure, in combination with residual welding stresses and, more significantly, very high thermal stresses arising during the original, post weld heat treatment (PWHT) were the factors instrumental in causing the defects. In addition, there was some evidence of hydrogen damage in the HAZ. While this may have provided some points of initiation for the subsequent stress relief cracking, it was concluded that it would not have been a necessary pre-condition for it.

In view of the above, it was considered necessary to devise specific weld repair and post weld heat treatment procedures which were not only code compliant, but also designed to avoid the re-occurrence of the problem. This was to be achieved by microstructural control of the weld repair HAZ and, where possible, manipulation of residual and thermal stress patterns.

This paper examines in detail how these procedures were developed and optimised and, in addition, outlines the development of the machining method for the repair weld preparation.

2. WELDING

2.1 Process and consumable selection

The initial information available in regard to the degree of defectiveness on the boilers showed that in general the defects extended from the external surface, or near surface, to variable depths through the shell thickness and contained over comparatively localised regions of the shell circumference. These factors suggested that, because of flexibility of access and ease of control in a non-uniform repair geometry, manual metal arc welding (MMA) would be the preferred process. Additionally the majority of field repairs discussed in the literature had also utilised this welding method. Although subsequent investigation showed the defects to be more widespread circumferentially, the MMA technique was retained and each 6/7 seam was excavated to a minimum 25mm depth.

The initial stages of development utilised a Babcock A2 (C-1.5% Mn) consumable conforming to AWS 5.1 E7018. Use of this lower yield stress consumable was intended to give a lower residual welding stress than would have been obtained with a matching consumable. Subsequent events, in particular the need to post weld heat treat at temperatures in excess of 650°C, rendered the A2 consumable unsuitable and it was decided to revert to the electrode type employed in the original fabrication, namely Babcock H1(Ni Cr Mo low alloy) confirming to AWS A5.5 E9018-M. This change required the production of electrode batches for development and the repair, since electrodes of the original H1 formulation were no longer available in the sizes required. It should be noted, from the weldability point of view, that A2 and H1 consumables, both of which are basic low hydrogen types, gave an identical welding performance and no parametric changes were required when adopting the latter.

2.2 Welding philosophy and joint design

The occurrence of stress relief cracking in weld heat affected zones is symptomatic of a lack of creep ductility during post weld heat treatment within a susceptible microstructure, namely the coarse grained HAZ. This was confirmed during the original defect characterisation where it was shown that the fine grained HAZ had a much lower defect incidence[1].

On this basis it was considered that a welding technique designed to produce a high level of grain refinement in the HAZ should be employed and the 'temper bead' method was thus adopted. Additionally, as the name suggests, this method of controlled deposition welding assists in softening of the HAZ which in turn helps to militate against the occurrence of hydrogen cracking. These factors were in addition to the code requirements of volumetric quality, strength, ductility, etc.

It was necessary to design the repair weld preparation to take account of the above factors, allowing access for effective control of the deposition sequence, but also ensuring that defective material from the circumferential welds was removed prior to repair. The original weld was of a double Vee configuration, shown in Fig. 2; the three welds in question, had been deposited by the manual metal arc process. The defects, in boilers 2C and 2D were associated with the outer part of these welds extending to a maximum depth of ≈27mm. In the case of 2A, the defective zone went to a depth of ≈35mm. Direct measurement of the outer weld width on the boiler surface showed that in some areas this approached 50mm. To ensure that defective material could be removed satisfactorily, the repair groove width at the plate surface was set at 70mm, thus removing a minimum of 10mm either side of the original weld metal and associated HAZs.

Fig. 3 gives details of the repair groove selected and is shown superimposed on the original weld preparation, in the same orientation as on the boilers. The repair groove depth was, for procedure development and qualification purposes, set at 35mm.

The side wall angles were 25°; this was considered optimum for the satisfactory deposition of the initial temper bead weld runs. As stated earlier, the depth of defectiveness on the boilers varied and in some localised areas widening of the preparation would be required. These factors were accommodated by introducing transition zones as shown in Fig. 4, thus allowing continuous welding of the initial layers into and through regions of variation in depth or width.

2.3 Material requirements

In the initial stages of weld parameter assessment, work was carried out using stock carbon/manganese plate material and on modern day BS1501-271B (Ducol W30). Some small quantities of original BW87A had been obtained, but this was used exclusively to support the comprehensive materials test programmes.

It was considered necessary, therefore, to procure material which matched, as closely as possible, the original boiler material specification in composition, mechanical properties and in microstructure. This was achieved in large degree by British Steel using an experimental basic electric furnace to produce ≈3 ton of liquid steel which was cast into slab ingots. These were then rolled to plate ≈58mm thick and normalised from 950°C. The time at temperature was chosen to generate a microstructure of similar grain size and constitution to that found in small scoop samples removed from several boiler plates (1).

In general, good agreement was obtained with respect to chemistry with satisfactory correlation on all the primary elements. On three of the four plates eventually produced, deliberate additions of As, Sn, Sb were also made but proved somewhat difficult to control with any precision. In mechanical property terms, a fairly wide range of tensile values was found for nominally identical compositions and heat treatment conditions, but all measurements were within the Magnox database bounding values for BW87A, being mean to upper bound. The microstructure and grain size correlated reasonably well with that found on boiler samples. The plates, in consequence, were considered to be an adequate representation of the boiler material and thus formed the backbone of the weld procedure development and qualification programme.

2.4 Test plate design

The test plates used for weld procedure qualification were required to replicate actual boiler conditions as far as possible. To this end, a simulation of the original circumferential weld was made in reproduction plate using the weld preparation in Fig. 2. The plates, each nominally 1000mm x 400mm, were assembled using large strongbacks initially (to simulate the constraint which would apply in practice) on the outer groove side. The reverse side was then welded fully using Babcock H1 electrodes. The test plate was set in the vertical position with weld metal deposited vertically up using a full weave for each pass. This was to introduce the maximum heat input possible and represent a 'worst case' as far as the original weld was concerned. After completion, this weld was dressed, a second set of strongbacks was attached and the first set removed. The first side weld was then backgrooved by manual grinding after which the weld was completed and the capping runs dressed off. After stress relief (with the second set of strongbacks still in place) at 600°C and post weld non-destructive testing (NDT), each test plate was subjected to a simulated service ageing treatment which used a step cooling sequence (1 day at 525°C, 6 days at 500°C, 21 days at 475°C), (2). This then was the 'starting point' for the repair procedure qualification test plate. The restraint provided by the strongbacks was such that there was minimal distortion typically ≤1mm deflection in both longitudinal and transverse directions.

2.5 Weld procedure development and optimisation
2.5.1 Development philosophy
There is a considerable volume of data published on the subject of the temper bead welding technique (3,6) and while this has involved materials other than the MnCrMoV (BW87A) steel used in the boiler shells, it formed a convenient starting point from which appropriate parameters could be developed.

The main objective of the development programme undertaken was to establish a workable set of welding parameters, robust and tolerant enough to apply to the excavation proposed for the defective areas of the boiler welds. The key areas addressed were:

- Design of the repair groove for weldability, metallurgical considerations and NDT requirements.

- Establishment of optimum weld run-out ratios/heat inputs for the temper bead layers.

- Selection of a bead sequence to facilitate welding of the temper bead layers within a repair groove.

- Establishment of a set of 'standard' welding parameters which were independent of welding position.

- The interaction of preparation features (radii at base and outer edge) with the welding technique in respect of HAZ grain refinement.

- The tolerance of the procedure to variations from optimum parameters.

The work of Allen et al (3) showed, for 2¼Cr1Mo material, that it was necessary to apply a buttering technique with controlled arc energy heat input in two layers around a repair excavation and, additionally, to specify a minimum weld bead overlap if consistent grain refinement of the HAZ was to be achieved. As a control measure for heat input, the use of run-out ratio R was advocated: where R is the ratio between the length of bead deposited and the length of electrode consumed. This was considered to be a better parameter for the welder to aim for rather than reliance on direct measurements such as current, voltage and welding speed. By achieving a specified run-out ratio, the welder automatically compensates for variations in current, voltage and travel speed and thus attains the desired heat input.

The repair programme welding trials therefore used R as one of the parameters set and monitored for the first four buttering layers of the repair weld deposit. For additional control, full monitoring of heat input was also carried out for each weld run in the first two layers through use of a PAMS (Portable Arc Monitoring System) unit.

A key component in the production of a fine grained HAZ is the relationship between the first and second buttering layers in terms of arc energy heat input. Dependent on other factors such as initial layer thickness and bead overlap, the ratio of second to first layer heat inputs (r) must fall between certain levels in order to ensure maximum grain refinement.

From the literature (3) relating to 2¼Cr1Mo material, a ratio of heat input of 1.78/1 was established. This was derived through direct calculation of heat input using electrode runout ratios of 1 : 1 in the first layer with a 2.5mm diameter electrode and the same ratio for the second layer, but with a 3.2mm diameter electrode. On this basis the same runout ratios were used on all the initial welding trials carried out at MBESL Tipton.

Another essential variable in satisfactory 'temper bead' welding is bead overlap. This is defined in Fig. 5; a bead overlap of 50% minimum was used for both first and second layers and was maintained for layers 3 and 4 in all of the initial trial work.

2.5.2 HAZ grain refinement evaluation
The degree of grain refinement of the weld HAZ was assessed metallographically. A transverse microsection was taken through the weldment, prepared to a 1μm diamond finish then etched in 2% Nital. The HAZ microstructure was then examined optically and prior austenitic grain size (PAGS) was measured. The dimensions of individual grains or clusters of grains with PAGS > 50 μm were then measured in a direction parallel to the weld fusion boundary. The remaining length of, by definition, fine grained regions along the length of the HAZ was then calculated as a percentage of the total HAZ length to give the HAZ grain refinement. For groove welds the whole associated HAZ was measured and used in the calculation, but for pads on flat plate the first and last bead HAZ's were discounted.

2.5.3 Welding trials and parameter selection
Initial trials comprised bead on plate welds on samples of mild steel plate set vertically with welding in the horizontal position. The purpose of this early work was to allow the welders to become familiar with the run-out ratio concept and accustomed to controlled bead overlap and provide as flat a weld top surface as possible. Heat input in the initial layers was very low, of the order of 0.5 KJ/mm, and this had implications with regard to adequate fusion and emphasised that each weld bead must be thoroughly deslagged, to maintain adequate weld quality. As stated previously, a run-out ratio of 1 : 1 was used for both layers with 2.5mm electrodes in the first layer and 3.2mm in the second.

Further tests on Ducol plate with the same parameters showed that there was insufficient heat input from the second layer to complete the grain refinement process. The second layer heat input was therefore increased by lowering the second layer run-out ratio to 0.7 : 1. Trials conducted at a preheat of 150°C established that this combination of first and second layer heat inputs resulted in grain refinement levels in excess of 90% with a bead overlap of 50% and a heat input ratio of ≈2.2 : 1.

After bead on plate trials, welds were made in grooved Ducol plate to establish an optimum groove geometry which would remove defective material, but be amenable to access for electrode manipulation and offer good weldability and vision into the joint. The requirements of subsequent NDT were also considered. The initial dimensions of the trial grooves were 25mm deep with a 20° side wall angle and 5mm radii at the bottom of the groove. These were used to develop the sequence of welding for the initial layers and the degree to which satisfactory grain refinement could be achieved. All groove welds were carried out with the groove horizontal and the plate vertical.

With regard to weld bead deposition sequence, this was initially as shown in Fig. 6. The first bead was deposited at the bottom external corner of the preparation and the layer was built up from that point, along the bottom edge, and round, until the full preparation was welded. With this sequence it proved difficult to obtain the optimum electrode angle onto the bottom edge and the welder was unable to see the area behind the weld beads and had difficulty in maintaining the correct bead overlap. The sequence was then altered to place the first bead into the bottom radius and then weld down the bottom face to the edge of the preparation. This allowed maintenance of the correct electrode angle. The layer was then completed by picking up on the first bead and working up the vertical face and out towards the top edge of the preparation.

Examination of these groove welds highlighted two other problem areas, at the radii on the bottom corners and at the outer edges at the mouth of the preparation. In both cases grain refinement to the level necessary was not achieved. At the corners the use of a 5mm radius gave an increased bead height; this was corrected by increasing the radius to 10mm. Welding at the outer edges of the preparation gave difficulty in holding a constant run-out ratio. Because of reduced heat flow in those locations, the tendency was to travel faster to avoid undue melting of parent plate. This problem was successfully addressed by incorporating run-out strips, of carbon steel, along the edges of the preparation, thus allowing the final weld beads to be deposited up from the parent plate and avoiding any inconsistency. The strips were attached by small TIG tacks which were subsequently incorporated in the first layer. On completion of welding, the run-out strip and associated overburden were removed by grinding. A general diagram of the finalised preparation showing the deposition sequence of the initial layers and including the run off strips is shown in Fig.7. It should be noted that identical welding parameters were used in each position within the groove. While this tended to increase the layer thickness, on the top face, the degree of grain refinement was not compromised.

With regard to grain refinement of the repair weld HAZ, this was achieved through the use of two weld layers as described above. The procedure developed included for a further two layers, essentially to the parameters of layer 2 (0.7 : 1 'R'; 50% bead overlap). The third layer produced a measure of tempering in the first layer HAZ; the fourth layer was deposited to preclude any alterations of the refined HAZ when filling out the remainder of the excavation with 4.0mm diam. electrodes. HAZ microhardness comparisons for first layer of weld and then for subsequent layers are given in Tables 1 and 2. This illustrates that peak HAZ hardness can be significantly reduced by the deposition of further layers.

While initial preheat used in the development of the technique was 150°C, there was concern about the possibility of hydrogen cracking especially since the first layer was deposited at very low heat input. This prompted the adoption of a minimum preheat temperature of 200°C. Additionally, in order to maintain low hydrogen (Scale D) levels for the H1 consumables, these were specified to be supplied vacuum packed for the repair contract.

Having established optimum welding parameters, it was necessary to tolerance these to evaluate the influence of variation in heat input in layers 1 and 2 and combine this with variable bead overlap. This was carried out on flat reproduction BW87A plate angled at 25°C to simulate the bottom face of the repair preparation. The results indicated a reasonable parameter tolerance, with +10% to –15% on first layer heat input combined with +6% to – 10% for the second layer. A 10% lowering of bead overlap in the first layer in conjunction with the above was also acceptable. These variations did not compromise the amount of grain refinement which remained at >95%.

2.6 Weld procedure qualification

The culmination of the welding programme was the requirement to carry out formal procedure qualification of the repair weld. Since the requirements of ASME XI and ASME III NB were to be achieved, as far as possible, weld procedure qualification was to ASME IX, but also satisfied those of BSEN 288-3.

The first welding development had taken place against an intention to carry out a non PWHT repair which then evolved into one utilising PWHT. Additionally, the proposed PWHT temperatures increased as development work was completed and finally the welding consumable was changed as outlined in Section 2.1. Thus, five procedure test plates were welded in total and a further two plates were welded to provide test material for stress relaxation, creep and creep crack growth testing within the materials programme (2).

The procedure test plates were prepared as outlined in Section 2.4 with strongbacks, a simulated 'original' weld PWHT at 600°C and a 'service ageing' treatment. The plates were then machined to form the repair groove which was standardised at a depth of 35mm. The plates were again set with the groove in a horizontal position and preheat was applied using electrical heaters to a minimum of 200°C for the final plate. Run off strips were attached and welding commenced as previously described. For the first two layers, each weld run was monitored by measuring the run-out length, and the electrode stub length. Welding parameters for each run were also recorded using a PAMS unit. Bead overlap was also checked on a regular basis through the use of callipers. This process was continuous for each bead in the first two layers and intermittent through layers 3 and 4. After welding of the first four layers, the remainder of the repair was completed using 4.0mm electrodes. This used a 'normal' no weaving fill sequence. Preheat and interpass temperatures were monitored by chart recorder.

After welding, the repair weld was dressed flush removing run off strips and overburden. The repair weld was then non destructively examined using magnetic particle inspection (MPI) and ultrasonic testing (UT). With strongbacks still attached, the weld was heat treated at the proposed (2) repair PWHT temperature of 675°C ± 10°C.

Procedure testing was then carried out to the requirements of ASME IX and BSEN 288-3 with, in addition, all weld tensiles at room and elevated temperatures. The results are shown in Table 3.

All of the test plates were subjected to comprehensive NDT and found to be free of significant defects. There was no evidence of hydrogen or stress relief cracking.

3. DEVELOPMENT OF A WELD PREPARATION MACHINING METHOD

With repairs to the boilers being carried out in situ, it was essential to provide a repair weld preparation method which could be applied directly to the boilers and be flexible enough to cater for variations in depth and width.

On this basis a portable machining method was proposed. This was based on a hydraulically powered mobile milling head which was mounted and ran on rails which would be attached round the boilers. The milling cutters used were designed to create a uniform cross section excavation on a single machining pass and a mild steel mock-up of a section of a Sizewell boiler was used for equipment commissioning, the development of optimum machining parameters, and an assessment of machine performance. Subsequently, the machine configuration and parameters were validated on a test plate of reproduction BW87A material. From the experience gained during these tests, various modifications were incorporated to improve machine performance and reliability.

3.1 Machining trials on boiler shell mock-up
The mock-up consisted of a piece of carbon steel boiler plate roll curved to an extrados with a 3486mm radius and all preliminary tests were carried out using this. For final validation, provision was made in the mock-up for the insertion of a section of BW87A to the same radius as the shell. This material matched the machining characteristics of the boiler material.

The first batch of cutters used were 60mm O/D with a 50° included angle, a 16mm flat nose and 10mm diameter radius. The maximum depth of cut was 36mm. Subsequent cutters for the final weld preparation were 70mm diameter and designed to accommodate depths up to 39mm.

During milling trials the following criteria were evaluated:

- Optimum speed of cutter rotation.

- Optimum feed rate.

- Maximum depth of cut safely attainable.

- The tendency of the machine to vibrate.

- The surface finish achieved.

- Accuracy of grooves produced.

- The efficiency of the compressed air/coolant spray for clearing and cooling the cutter.

3.2 Assessment of results

The results from these trials were encouraging. It was found that about 100 rpm was the best cutter rotation speed and a full cut depth could be achieved in a single pass at a feed rate of 10 mm/min. Initially, there were some problems with machine vibration which had implications for the surface finish but, after the machine had been modified to stiffen it, 3 μm Ra or better was attained. The finished groove was accurate to within 1mm and the compressed air/coolant spray was effective, giving good cutter life at optimum speed. The trials were carried out on carbon steel and BW87A plate with the latter giving a better cutting performance.

The test work covered the full scope of use of the machine for the 6/7 circumferential seams of boilers 2A, 2C and 2D which were to be repair welded and the 2C 5/6 seam which was required to be profiled. When the machine was operating in the final stages of these trials at maximum cut depth, it was performing well within its capabilities.

The requirement to accommodate different depths of cut within any one repair excavation, was achieved by step milling 1 in 5 slopes between depth transitions. These had to be hand finished with grinding to remove steps and corners. A similar approach was used where localised widening of the repair excavation was required.

Prior to deployment at Sizewell, the milling machine was shown to be capable of generating repair weld preparations to the required specification. It was also demonstrated that it could be set up and operated in a reliable, repeatable and safe manner and adequate experience was gained to ensure that a successful machining operation could be performed at site.

4.   POST WELD HEAT TREATMENT PROCEDURE

Clearly PWHT of the repaired weld would be a major undertaking. Normally PWHT of structures such as the boiler shell is carried out when they are essentially plain cylinders and although all attachments are in place they are of minimum size and thermal gradients can be simplified. In the case of the weld repairs the boilers were anything but plain shells, were located in amongst other plant and contained internal structures. Furthermore, the rigorous safety case requirements, due to the absence of a proof pressure test, necessitated detailed knowledge of temperature and thermal stress transients for all affected parts of the boiler, not just the repaired weld. This, in turn, required detailed monitoring, and control, of the temperature profile.

Arising from the metallurgical investigation of the original cracking (1) another constraint on the PWHT was a need to raise the temperature as fast as possible, subject to the ASME limit of 98C°/hr in order to minimise the time spent at temperatures where susceptibility to stress relief cracking was greatest.

The PWHT operation was required to have a very high level of reliability because the normal options of 'holding' the whole PWHT at the transient condition when a fault occurs while it is rectified, or aborting the whole PWHT, could substantially increase the likelihood of stress relief cracking. Continuing with a fault condition, say a heater failure, was also unlikely to be acceptable due to the locally high thermal stresses which would be generated. Put simply, failure of a PWHT operation would very likely be irrecoverable and render the boiler, and hence the reactor, unfit for future service.

Accordingly, it was decided that a substantial specification development and optimisation programme was necessary to develop a high level of confidence that the PWHT would reliably and repeatedly attain its objectives. The programme comprised the following:

- Specification of PWHT parameters.

- Selection and description of PWHT heating method.

- Determination of power system requirements.

- Thermal insulation of internal and external surfaces.

- Preheat arrangements and the transition to PWHT.

- Organisation and layout of heater power supplies, monitoring and control instrumentation.

- Rig trials.

- Venting and cooling system.

- Contingency planning for fault conditions.

4.1 Specification of PWHT Parameters
4.1.1 Code Requirements
A number of documents provide guidance on the method and extent of heat treatment required in this instance. It was found that several documents gave similar advice with one of the most comprehensive being BS5500: 1997, Section 4.4 (7). The BS5500 requirements for a fully circumferential heated band are that:

- the weld and its HAZ attain the specified PWHT temperature.

- the heated band width should be at least $5\sqrt{Re}$ wide (in this case 2200mm) where R and e are the vessel radius and wall thickness respectively.

- the temperature at the edge of the heated band shall be at least half the temperature at the weld.

- the vessel adjacent to the heated band shall be insulated to prevent harmful thermal gradients.

- the maximum rate of heating is 6000/e °C/hr, in this case 104°C/hr. However the maximum heating rate allowed under the ASME code is 98°C/hr and this was used as the limiting value for the PWHT.

4.1.2 Finite Element Model.

Finite element analyses were carried out to determine the thermal stresses during various stages of the heat treatment cycle. The repair weld seam 6/7 is located approximately 2413 mm from the dome head as shown in Fig.1 Approximately 398 mm beneath the repair weld 6/7, in course 6, is a set of thermal sleeves through which the tubes feeding the HP superheater inlet bank pass. The exit from the HP superheater is near the top of course 7 in a cooler region of the shell during the heat treatment and is therefore of less concern as allowable stresses are considerably higher. Further beneath the 6/7 weld at a distance of 1319 mm are a further set of thermal sleeves through which the tubes from the HP boiler outlet bank pass. These are also in a cool region during the heat treatment process and again of less concern.

There were two periods during the heat treatment process which were immediately identified as a cause for concern. The first is during preheat when, due to the presence of the weld preparation there is a localised thin ligament which should not be overstressed. Secondly, when the maximum PWHT temperature is applied to the completed weld repair at the end of the heating cycle and maximum thermal stresses are generated in the weld region.

The numerical representation of the preheat and loading to the post weld heat treatment temperature used an axisymmetric finite element model, Fig. 8. The mesh which was generated using the graphics pre and post processing package PATRAN (8), was purposely kept relatively simple. The mesh comprised a cylindrical section and a dome head. The most complex feature of the mesh was the weld preparation at weld seam 6/7 and the profiled groove at seam weld 5/6 where all geometric features were modelled in detail.

The temperature loading during the heat treatment of the boiler shell is from heating pads fixed to the outer surface of the shell. In the finite element model, nodes on the outer surface over the length of the heated area were prescribed temperature values matching those expected from the heating pads. The inside surface of the shell was divided into zones. A heat transfer coefficient and an inner surface air temperature were calculated and prescribed for each zone.

The derivation of the complete nodal temperature distribution and subsequent stress distribution was carried out in two stages using the numerical solver ABAQUS (9). A temperature transient analysis comprising of steady state operation at a preheat temperature of 200°C , followed by a heating phase load cycle to a maximum temperature of 685°C and finally a period at the steady state post weld heat treatment temperature was applied. This was followed by a thermal stress analysis, where the nodal temperatures from the previous analysis were read into the ABAQUS input file and the elastic stresses calculated at each time step.

4.1.3 Preheat

During the preheat operation the heating pads are removed from the weld area to allow welder access. It was essential that stresses induced in this area and especially in the remaining weld ligament should remain low. The analysis indicates that the areas of high stress are at the base of the weld excavation and the end of the heat treatment area, particularly at the 5/6 groove. However, the level of stress is relatively low compared to the yield strength of the material and is therefore considered to be acceptable. Higher tensile stresses were generated at the groove in seam weld 5/6. The level of stress was considerably higher than at weld 6/7, but as the 5/6 seam weld area experiences a relatively low temperature during the heat treatment, the level of stress when compared against the allowable stress for the material at temperature was shown to be acceptable.

4.1.4 Post Weld Heat Treatment.

The model showed that for the BS5500 minimum heated band width and temperature distribution the thermal stress at the repair weld at the PWHT temperature was significantly in excess of the yield stress. Alternative arrangements were assessed using the model with the objective of reducing thermal stress to below the yield stress. This was almost completely achieved by using a heater band width of 4400mm with the half temperature position at 1800mm from the weld. This arrangement, however, gave rise to high stresses at the intersection between the cylindrical main shell and the hemispherical top dome. The solution was to extend the insulation onto the dome, but heat loss from the dome was necessary to prevent overheating of internal structures. The final compromise was to extend the insulation 1100mm up onto the dome.

To limit the time spent in the temperature range where the material creep ductility is low, it was proposed to raise the boiler shell temperature at the maximum possible ramp rate commensurate with ensuring that the thermal stresses generated do not cause damage to the shell. There were two options for the axial temperature distribution during the heating phase. One was to raise the temperature across the whole heated band at a constant rate and putting the outer bands into a 'hold' condition when they reached their intended temperature for the proposed axial temperature distribution during the soak period. The other was to heat all bands at differing rates such that they all reached their final intended temperature at the same time. In order to exercise the greatest control over heating process and to limit thermal stresses the latter method was proposed and then assessed for heating rates of 70°C/hr and 100°C/hr at the weld

The stress state was examined at numerous time steps during the heating transient. As expected the most onerous stresses occurred at the end of the heating cycle when steady state heat treatment commenced. Comparison of the stress level at the end of the heating cycle showed that the faster 100°C/hr heating rate generated the higher stress. However, the difference between the 100°C/hr ramp rate and 70°C/hr was only of the order of 5 MPa. The slight increase in stress was outweighed by other benefits, and it was therefore decided to use the faster heating rate. As the stress level was less than yield magnitude and met all other criteria, the heating profile was judged acceptable.

4.1.5 Thermal Sleeves

Although the overall temperature and stress profile was acceptable, the thermal sleeves associated with the HP Superheater Inlet needed to be checked in greater detail. To carry out this activity it was necessary to generate a 3D, 20 noded brick model. In order to work within the computing resources and facilities available the model was restricted to a 10 degree segment of the shell. This model included the completed repair weld at 6/7 , four thermal sleeves and the groove at the 5/6 seam weld.

In practice small heating pads will be attached around the thermal sleeves. Thus the linearly varying temperature distribution prescribed in the axisymmetric model for the outer surface of the shell is still applicable.

Examination of the temperature distribution at the thermal sleeves showed that a drop in temperature of approximately 60°C could occur between the outer surface and the innermost point of the sleeve. The relatively high temperature differential generated peak stresses at the toe of the thermal sleeves of 158 MPa acting in the axial direction. Although this stress was high, it was very localised and still below yield magnitude.

4.1.6 Sensitivity Study on thermal sleeves.
Although there was a great deal of confidence in being able to control the heating pads and that the method to be employed for the heating process was achievable, due to the complexity of the process and the uncertainty of input parameters used in the finite element analysis a sensitivity study was carried out.

(a) The first case to be analysed examined the effect of increasing the temperature tolerance band of the heating pads. The heating pads chosen operate with a tolerance limit of ±10°C on the mean band temperature. This means there could be a temperature gradient between the bottom and top of the thermal sleeve of 20°C. In the finite element analysis this was increased to ±20°C for the pads on the outer surface of the sleeves.

The thermal sleeves pass through the shell and are fillet welded to the inner and outer surfaces. There is effectively an area of unfused land between the two fillet welds over the thickness of the shell. In this worst case scenario it was assumed that there would be no heat transfer between the shell and thermal sleeve. All heat flowed through the fillet welds.

As a consequence of this worst case set up, peak axial stresses at the toe of the thermal sleeve fillet weld increased to 279 MPa.

(b) The next stage involved reducing the temperature tolerance across the thermal sleeves to ±10°C. However, it was found that this had very little effect on the stress induced at the weld toe, presumably because of the heat transfer path.

(c) It was then decided to model a more realistic heat flow between the thermal sleeve and the boiler shell. Due to radiation effects it was considered feasible that the temperature across the unfused land would equalise and in effect there would be full heat flow across the junction.

When modelled, stresses at the toe of the weld decreased to 265 MPa acting in the axial direction. Examination of the effective stress indicated that yielding could occur at the weld toe at maximum temperature conditions. The area of high stress extend over a very small surface area and was only 2 mm deep. It was also judged that the finite element model was pessimistic in that sharp corners had been modelled at the weld toe.

(d) Further investigations examined variations in temperature applied to individual thermal sleeves such as one thermal sleeve temperature being at the upper limits of the temperature band whilst the adjacent sleeve at the lower limit.

It was, however, generally found that the maximum stress at the weld toe did not vary significantly and that it was driven by the individual temperature of the sleeve relative to the shell band temperature, rather than effects from the adjacent sleeve.

From the linearly varying temperature distribution and the upper bound distribution imposed during the sensitivity study it was felt that the possible stress distributions had been adequately bounded.

4.2 Heating method

Given the location of the boiler, surrounded by other plant, and the need for fine control of temperatures in areas of complex geometry, electrical heating was considered to be the only viable method. The two most suitable electrical heating methods are induction heating and resistance heating. The former was used for the original local PWHT of the circumferential butt welds during construction but the latter was chosen for the repair PWHT on the basis of greater operational experience and perceived better control capabilities.

Individual heating elements comprise a stranded resistance wire running through ceramic blocks. The purpose of the blocks is to provide electrical insulation but their design allows the element to have good flexibility to conform to the surface being heated.

Standard $\approx 0.05 m^2$ heater elements are powered by a 110V supply and are rated at 4 kW. Within these parameters various profiles, mainly rectangular, can be made to allow the element layout to be customised to the component being heated. In addition, there is scope for wiring heater elements in parallel or series or using lower supply voltages for smaller specialised heater elements.

One or more heater elements can be fed from a single power supply and as such are referred to as a single heater. Each heater has its own control thermocouple, a type k chromel-alumel thermocouple with each wire capacitance discharge welded to the component being heated. The attachment point is normally at the geometric centre of the heater.

Each thermocouple is connected to a dedicated channel of a multi-channel (typically 6 channels) controller unit. The major parameters of the heat treatment can be programmed into the controller units - start temperature, heating rate, soak temperature, soak time, cooling rate and final hold temperature. By comparing the temperature indicated by the control thermocouple for each heater with the programme, the controller switches a high current relay to control the power supply to that heater. Thus the controller unit will ensure that each heater will be independently controlled to achieve the programmed temperature profile.

Normally the control thermocouple is connected in parallel to a chart recorder which provides a continuous record of temperature and basis for monitoring the progress. There have been reports that using the same thermocouple for both control and monitoring purposes can introduce errors. Although the problem was thought to only apply to earlier generations of low impedance equipment some simple trials were carried out. While they conferred that there was not an impedance problem they did reveal an error due to the open circuit detection (OCD) on some controllers. The OCD, which injects a small voltage through the thermocouple to ensure it has continuity, was set high and if the recorder sampled the thermocouple at the same time, it would detect the combined emf resulting in an incorrect reading. As a precaution, even though all of the equipment was corrected in this respect, and shown to be error free, independent monitoring thermocouples were installed adjacent to every control thermocouple and connected directly to the recorders.

4.3 Determination of power system requirements
In order to calculate the power requirements it was necessary to consider:

- Thermal capacity of the shell.
- Axial conduction along the shell to the zone beyond the heated band.
- Radiation and convection losses from the outside surfaces, i.e. from the insulation laid over the heaters.
- Radiation and convection losses from the inner surfaces of the shell, influenced by:
  - efficiency of internal insulation.
  - internal air convection current behaviour.
  - thermal capacity of internal structures.
  - effect of a boiler internal cooling system, if fitted.
  - conduction along connecting the thermal sleeves to the tube banks.
- Efficiency of heaters:
  - Individually.
  - as a whole system - only some heaters will have a high duty cycle, others will operate at relatively low utilisation.
- Heating rates - the power requirement will increase with heating rate and maximum power will be required just before the maximum temperature is attained.

There are uncertainties, of differing magnitudes, associated with all of the above, and this combined with the fact that at no point could the PWHT be considered to have attained a steady state made it difficult to predict the power requirements with any certainty. Thus estimates varied from < 0.5 MW for a perfectly insulated shell to approximately 10 MW if every heater was working at 100% duty cycle. Logistic and programme requirements meant that an early decision had to be made and considerable engineering judgement was involved to arrive at a requirement of 5 MW for the 415v supply.

4.3.1 Heater power supplies
As mentioned earlier, each heater is powered by a low voltage supply which is obtained from an outlet tapping on a 415V transformer. Although the number of heater elements could be reasonably predicted from the knowledge of the area to be heated, the number of heaters required less clear. The initial estimate of the number of heaters, and hence the number of transformers, was based on a subjective assessment of acceptable heater sizes, for control and to accommodate local features with varying heat losses. The uncertainty in this initial estimate was one of the major driving forces for carrying out the rig tests described later.

4.4 Thermal insulation
On the external surface operatives would be in close proximity to the boiler and the need to

control the environment in the relatively small boiler cell, meant that irrespective of heater power/heat loss considerations, insulation was mandatory. On the inside surface these constraints did not necessarily apply. However, as the reviews progressed, it became clear that excessive dome and tube bank support structure temperatures, or a massive (maybe 1 MW) load for a cooling system, combined with unacceptably high through-wall temperature gradients, meant that insulation of the internal surface of the shell was also mandatory.

4.4.1 Internal surfaces

Two of the design features of a boiler in a gas cooled reactor system are that:

- the hot gas is constrained to flow through the tube banks.

- the dimensions of the boiler pressure vessel are kept to a minimum.

The result is that where the tube banks abut the shell internal surfaces the clearances are small, ≈ 25-40 mm, and in these and other areas there is extensive baffling attached to the shell. It was intended to remove the bulk of the baffles but, apart from a vertical lift to unload the supports, the tube bank to shell clearance could not be improved. Where there was access to the internal surface of the shell, insulation was relatively straightforward comprising layers of fibrous insulation (Superwool 607) attached to the shell using capacitance discharge welded studs and speed washers. The particular challenge was to insulate the internal circumference, which is about 12m in length, where the tube banks allowed only minimal access. Post repair operation considerations meant that any insulation installed in these areas would have to be thermally stable and completely removed before return to service. This ruled out any form of sprayed insulation or use of particulate insulation. The solution finally adopted was the use of high efficiency microtherm insulation attached to stainless steel sheets which were slid around the gap between the tubes and the shell surface.

Extensive trials were carried out to optimise this system which was particularly difficult to install as the access constraints meant that the long panels (up to 3.9m) had to be riveted together in-situ as they were fed in from small ≈ 600mm lengths. The system finally evolved was very effective and robust in that it used the horizontal baffles and their attachment bolts to guide the panels into position and the natural spring of the stainless steel sheets reacting against the curve of the shell held the insulation in contact with the shell surface. A fundamentally similar system was used to insulate the horizontal baffles.

4.5 Preheat arrangements and the transition to PWHT

It was decided that once welding commenced the repair would not be cooled to ambient temperature until the completion of PWHT which gave some to some particular equipment and operational requirements. It was intended that the shell would be preheated up to the edge of the weld preparation, but that during welding and subsequent hot NDT heaters over a limited area would be removed for access. Theoretical analysis predicted that sufficient heaters could be removed to give good access for their operations without giving rise to unacceptable temperature reductions or thermal stresses.

In order to facilitate repeated heater installation/removal in this area modified heaters with integral insulation and a steel backing strip were used close to the weld. They were held in place by a simple framework mounted on studs welded to the shell. This system was validated during the rig trials. These heaters could not be used to attain the PWHT temperatures and heating rates and were replaced by 'permanent' heaters immediately prior to PWHT. Appropriate allowances were made in the overall heater layout for the preheat heaters and their subsequent replacement by the permanent heaters to obtain the necessary continuity of heaters at all times. Detailed method statements and quality plans for the temporary removal and replacement of preheat heaters and their final replacement were prepared. This documentation included the alterations to the wiring and monitoring arrangements which had to cope with the added complication that the permanent heater layout differed from that used during preheat.

4.6 Organisation and layout of heaters and the control and monitoring system

Ideally a heat treatment will operate largely under automatic control and the operators are only required to monitor temperatures and make minor adjustments to the controls. However, if major deviations occur their effects can often be mitigated by adjusting adjacent heaters to the limits of their tolerance band or, in the case of component failure, a back-up system or replacement component can be utilised. For this to occur expediently a logical, clearly laid out, system is essential and considerable effort was expended in the development stage to put this in place.

In order to achieve the basic temperature profile a series of bands of heaters were placed on the shell, ideally all heaters within a band would be controlled to the same temperature and individual bands would be controlled to different temperatures to achieve the desired axial temperature gradient. The height of each band of heaters was optimised to maximise the number of complete bands and where internal or external attachments (particularly thermal sleeves) occurred they would coincide with a particular band. As far as possible all the heater controllers, their power transformers and their monitoring thermocouples for a particular band were grouped together so that any unusual behaviour could be quickly detected. All cables were clearly identified, routed as a group and connected to the appropriate equipment according to a clearly designated layout. Fig. 9 shows some of the features of the heater and thermocouple layout.

4.6.1 Control and monitoring system

The system of programmed controllers with operators monitoring chart recorders and making minor adjustments was considered to be only marginally adequate for such a complex heat treatment with high heating rates. For example, for a complex band it could take over 10 minutes to record the individual temperatures and calculate average, minimum and maximum values for comparison with the required mean and bounding temperatures. Allow a few minutes to make adjustments and the band mean temperature would have increased to $\approx 20°C$ above that at the start of the scan and it would be time to restart the process. This would impose a dangerously high workload on the operators with the attendant possibility of mistakes, and make it difficult to identify trends before they developed into out of specification conditions.

Accordingly, it was decided to use a PC based system to assist in the real time analysis of the temperature data. Consideration had been given to using a PC system to replace all of the chart recorder functions but the demands of developing sufficient software and hardware robustness were considered too onerous, given that this was a critical application. Using the digital data output facility on the chart recorders, a PC would periodically scan every monitoring thermocouple. It would calculate the mean, maximum and minimum temperatures for the bands, compare them with the programmed values and identify all thermocouples which exceeded either the warning limits (arbitrarily set at 5°C within the specified limits) or the specified limits themselves.

The summary output from the PC was a single sheet summary confirming the status of each individual band, highlighting warning, or out of specification, situations and, less frequently, a printed summary of all thermocouples segregated by band, again highlighting individual warning or out of specification conditions. Thus a rapid scan of the output sheets allowed the important data to be clearly identified and corrective measures implemented. Normally the PC scans were carried out every 20 minutes, but additional scans could be carried out on demand if required.

4.7 Rig trials
Clearly the uncertainties and need for optimisation meant that it was inadvisable to undertake a PWHT on the plant without having carried out realistic trials. A full scale rig reproducing one third of the circumference of the boiler, ≈3m high and containing one half of the bank of thermal sleeves closest to the repair weld was constructed.

The simulated position of the repair weld was near the top of the rig which, neglecting peripheral heaters to cancel out edge effects, allowed 1/6 of the proposed PWHT to be simulated which, given symmetry, was considered adequate.

The initial trial utilised the best-estimate of heater and control layout and incidentally evaluated the internal and external insulation proposals. These arrangements were modified for subsequent trials which also included simulation of significant internal attachments and additional thermocouples to monitor short range temperature profiles between the planned monitoring thermocouple positions in the vicinity of these attachments. The trials were also carried out at increasing heating rates until rates of 100°C/hr could be reliably attained.

The main benefits of carrying out the rig trials were:

- Optimisation of the heater layout and monitoring arrangements to attain the desired heating rate and temperature distribution both on the plain shell and in the vicinity of attachments.

- Confirmation that unexpected short range temperature gradients did not exist, or could be eliminated by revised heater layouts.

- Confirmation that through-wall gradients were acceptable providing the internal insulation was sound and, conversely, confirming that deficiencies in the internal insulation, even in small areas could lead to undesirable through-wall gradients.

- Confidence in the robustness of the heaters, and the control and monitoring equipment.

- Demonstration of the value of the PC system and the opportunity to optimise the operation and information flow to maximise its usefulness.

- The opportunity to refine the cabling layout and identification proposals in the light of practical experience.

- The opportunity to develop and refine the control and monitoring methodology.

- Demonstrate the robustness of the PWHT proposals and give confidence to stakeholders and independent bodies.

4.8 Venting and cooling system

When gas circuits are opened for access they are held at negative pressure and the exhaust from the fans is filtered and monitored to ensure that any radioactive particulate matter or gaseous species are either contained or released in accordance with statutory limitations. Clearly when the boilers are heated there will be expansion of the air within the boiler and the potential for release of species which had condensed onto, or been absorbed into, internal surfaces. Thus there was a need to maintain the controlled ventilation process during PWHT, though precautions would have to be taken to limit the filter air inlet temperature.

Cooling of the internal structures was not so readily resolved. As has already been stated the internal heat losses were difficult to calculate. However, the rig trials indicated that the heat losses from the internal surfaces would not be excessive and it was likely that internal overheating would not occur. Nevertheless, because overheating could not be ruled out it was decided to install a cooling system on a contingency basis.

Direct cooling of the boiler tubes with air or water was an obvious possibility but, apart from the fact that the external connections had been removed, the thermal stresses at the hottest bank of thermal sleeves made this option unattractive. Use of refrigeration plant to supply cold air to the boiler, or to cool extracted hot air, increased the power supply problems and in the latter case gave rise to a contamination problem in the chillers.

The solution finally adopted was to extract air at the end of the gas inlet duct via multiple fan-filter units which draw air from a manhole at the base of the boiler and another manhole in the top duct adjacent to the top of the boiler. The former gave an airflow through the boiler to provide cooling and the latter provided tempering air which cooled the hot air from the boiler to an acceptable temperature before it reached the filters.

Variable dampers were fitted at the inlet points which, together with dampers on the fan inlets and the facility to vary the number of fans in operation gave a very flexible system. Prior to the PWHT a test programme confirmed the combinations of damper settings which formed the basis of the operating regime.

4.9 Contingency planning for fault conditions

As early termination of the PWHT or operation with a major fault was not acceptable contingency plans to deal with system failures were made in advance.

- There was redundancy in the main electrical supply system with the facility to switch supplies from distribution boards designated for other boilers even if that impinged on the preheat and welding operations.

- Method statements were prepared for alterations to the electrical supply and a team of qualified personnel were trained.

- Spare 415v transformers, heater controllers, high current relay units and chart recorders were provided with leads and connectors to facilitate speedy substitution.

- All control and monitoring thermocouples were duplicated and spare heater power cables and thermocouple compensating cables were run between the boiler and the control area.

- It was judged that individual heaters could be replaced or doubled up with a new heater, but in critical areas such as the hot zone and inaccessible internal components, heaters were duplicated.

- A thermal imaging camera was employed to continuously monitor for connection problems or cable overheating and identify them before they became a major problem.

- Finally a lead team of individuals familiar with the design and operation of the PWHT operation, thermal stress calculations and metallurgical considerations were identified to be present throughout the PWHT in order to make an informed judgement on how to deal with any contingency which arose.

5. CONCLUSION

An extensive and rigorous specification, development and optimisation programme was carried out. It started with a theoretical consideration of the requirements of the welding and PWHT operations which were then developed in experimental trials and simulations. It culminated in detailed specifications which were well founded and shown to be attainable. The programme gave confidence that the repair operation had a high probability of success and there were no hidden threats remaining to be resolved during the implementation phase which is described in reference 10.

6. REFERENCES

1. Exworthy L F.. Flewitt P E J, and Ellis BJC. Paper 2, these proceedings.

2. Hunter A N R. McDonald E J. Moskivic R, and Lamb M. Paper 8, these proceedings.

3. Allen D, Kelly T: "Cold Weld Repair – Development and Application", 2nd International

EPRI Conference, May 1996, Daytona Beach, USA. EPRI 1996.

4. Freidman L M: "Repair Welding for High Temperature Equipment in Power Generation and Refinery Service", 10th Annual North American Welding Research Conference, October 1994.

5. Freidman L M: "EWI/TWI Controlled Deposition Repair Welding Procedure for 1¼Cr½Mo and 2½Cr1Mo Steels", Pressure Vessel Research Council/EWI Workshop, Jan/Feb 1996, San Diego, USA.

6. Jones, R L: "Development of Two Layer Deposition Techniques for the Manual Metal Arc Repair Welding of Thick C-Mn Plate Without Post Weld Heat Treatment", The Welding Institute Research Report 335.1987, April 1987.

7. BS5500:1997. Specification For Unfired Fusion Welded Pressure Vessels. BSI 1997.

8. PATRAN. The MacNeal – Schwendler Corporation. September 1998.

9. ABAQUS. Users Guide. Hibbitt, Karlson and Sorenson Inc.

10. Evans H V. McDonald E J and Wilkens A. Paper 9, these proceedings.

7.   ACKNOWLEDGEMENT

This paper is published with the permission of the Director Technology and Central Engineering, BNFL Magnox Generation.

The welding development and procedure qualifications trials were carried out at MBESL Tipton under the guidance of J Tolaini.

Dr P Jeans and Dr G Little gave valuable advice on thermal stresses

D Cottrell, R Jagger and A Taylor bore the brunt of the development of the PWHT procedure and technicians from Didcot Heat Treatment made valuable contributions to the system design as well as operating the trials under the guidance of A Anderson.

CME Ltd. developed the communication system and software for the temperature monitoring PC system.

Table 1: Micro Hardness Survey (500g) - Test Plate Ref W38.

Single layer weld carried out using 2.5mm dia. electrode at 150°C preheat on reproduction BW87A - as welded. (sample ref. B)

Fusion line	Hardness	Fusion line + 1mm	Hardness
From	301		312
to	423		411
ave	362		358

Table 2: Micro Hardness Survey (500g) - Test Plate Ref TP14.

Completed weld carried out using 4 layer, temper bead technique at 150°C preheat on reproduction BW87A - as welded. (sample ref. C)

Fusion line	Hardness	Fusion+1	Hardness	Fusion +2	Hardness
From	290		285	25	266
to	330		330	26	312
ave	315		305	27	289

TENSILE TESTS

Specimen Location	Test temp °C	Area/mm^2	Ultimate Load kN	Ultimate Stress N mm^2	Failure Location
Cross weld 1	Room	1092.9	572.9	524	Parent plate fracture
Cross weld 2	Room	1082.7	570.9	527	Parent plate fracture

ALL WELD TENSILE
LONGITUDINAL - H1 Weld Metal

Test Temp °C	CSA/mm^2	RP2/N mm^{-2}	Load/k N	UTS/N mm^{-2}	Red. of Area/%	Elongation/%
Room	125.9	496	73.40	583	74	31.5
360	126.3	406	65.96	522	69	28

GUIDED BEND TESTS

Type (D=4t, 180°)	Result
Side Bend	Acceptable
Side Bend	Acceptable
Side Bend	Acceptable
Side Bend	Acceptable

Hardness Survey
Vickers Load=10Kg

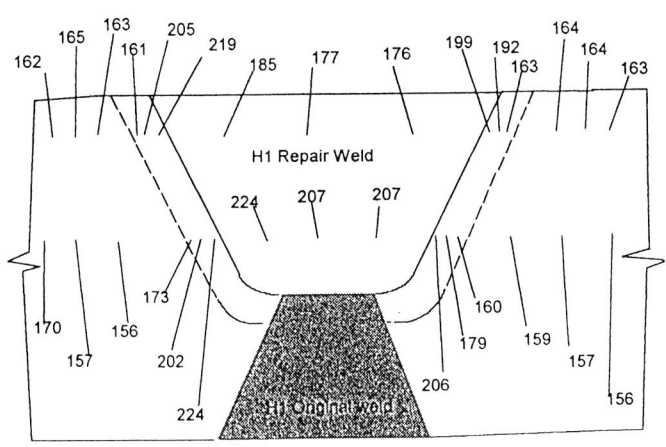

Table 3: Mechanical Test Results.

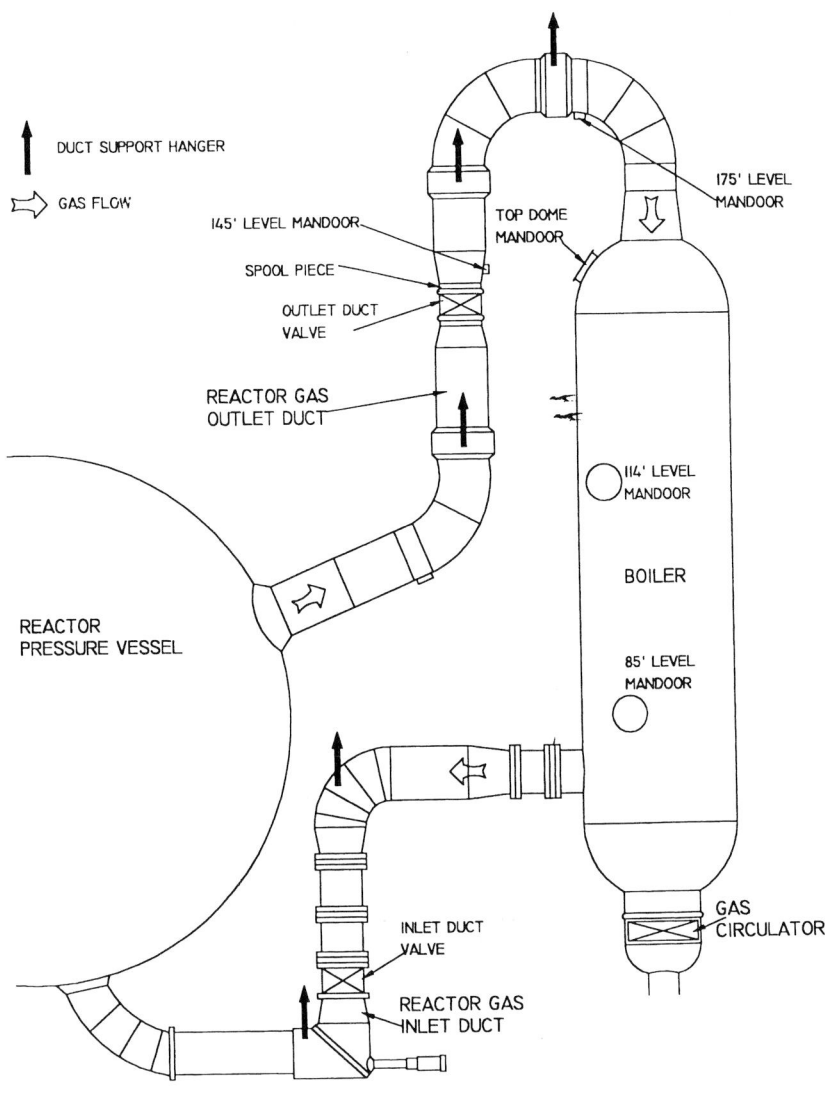

Figure 1: Schematic layout of gas circuits at Sizewell A

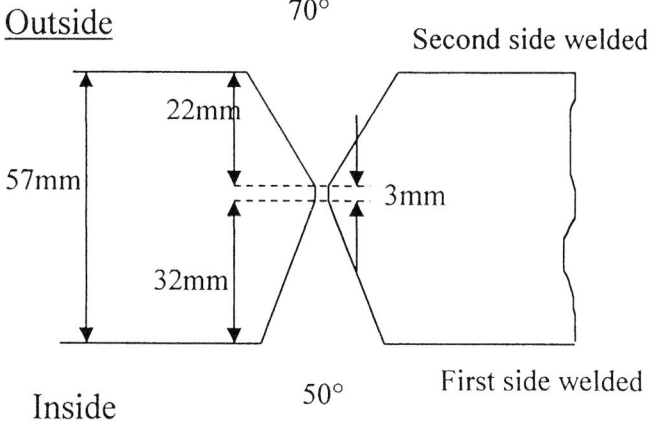

Figure 2: Original Weld Preparation.

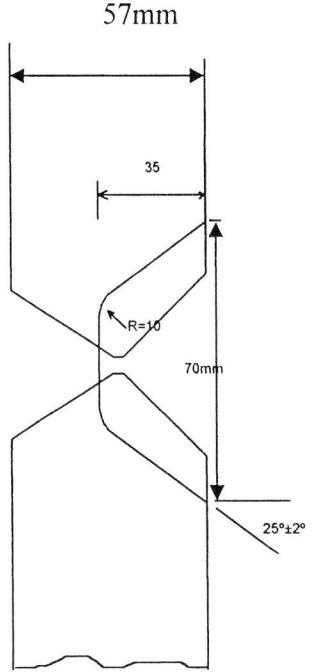

Figure 3: Developed Repair Profile.

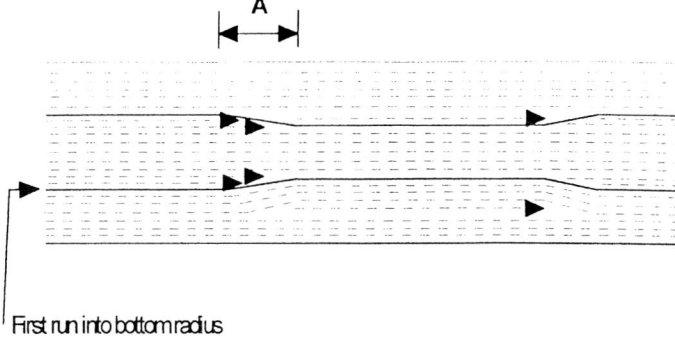

First run into bottom radius

Dimension A : 50mm min for a transition height of 10mm

Figure 4: Indicative Run Sequences for Areas of Groove Depth Transition.

<u>Measurement of Bead Overlap</u>

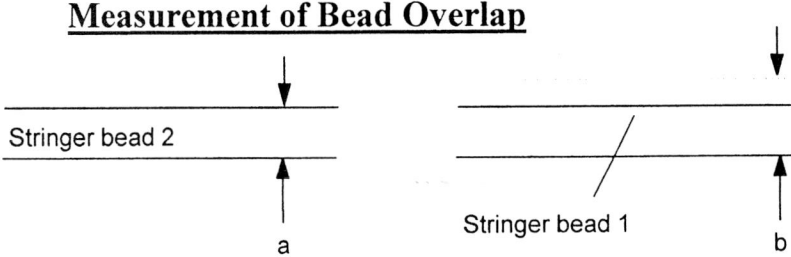

Bead overlap is been defined as :

% Bead overlap = [a - (b-a)] x 100 / a

For example if a is 6mm and b is 9.5mm, the % bead overlap is :

[6 - (9.5 - 6)] x 100 / 6 = 42%

Figure 5: Measurement of Bead Overlap.

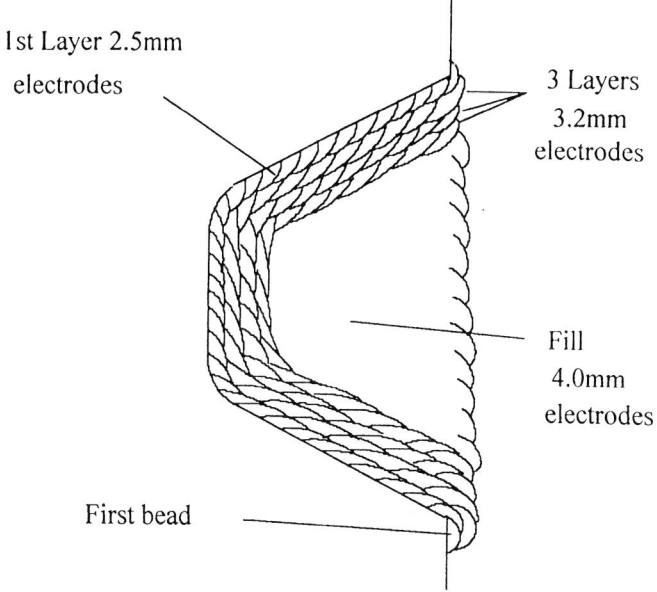

Figure 6: Initial Weld Bead Deposition Sequence.

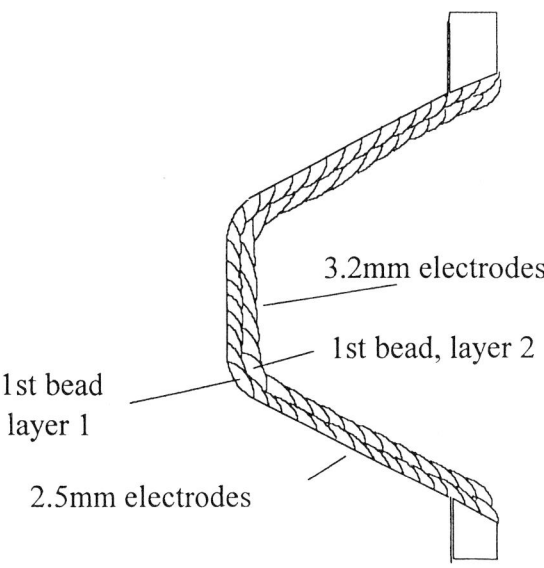

Figure 7: Weld Bead Deposition Sequenceof the Initial Layers.

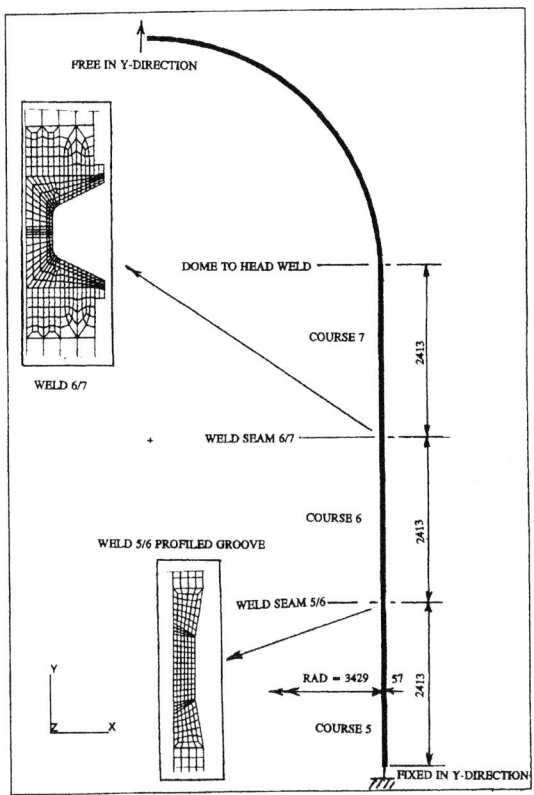

Figure 8: Axisymmetric Finite Element Mesh.

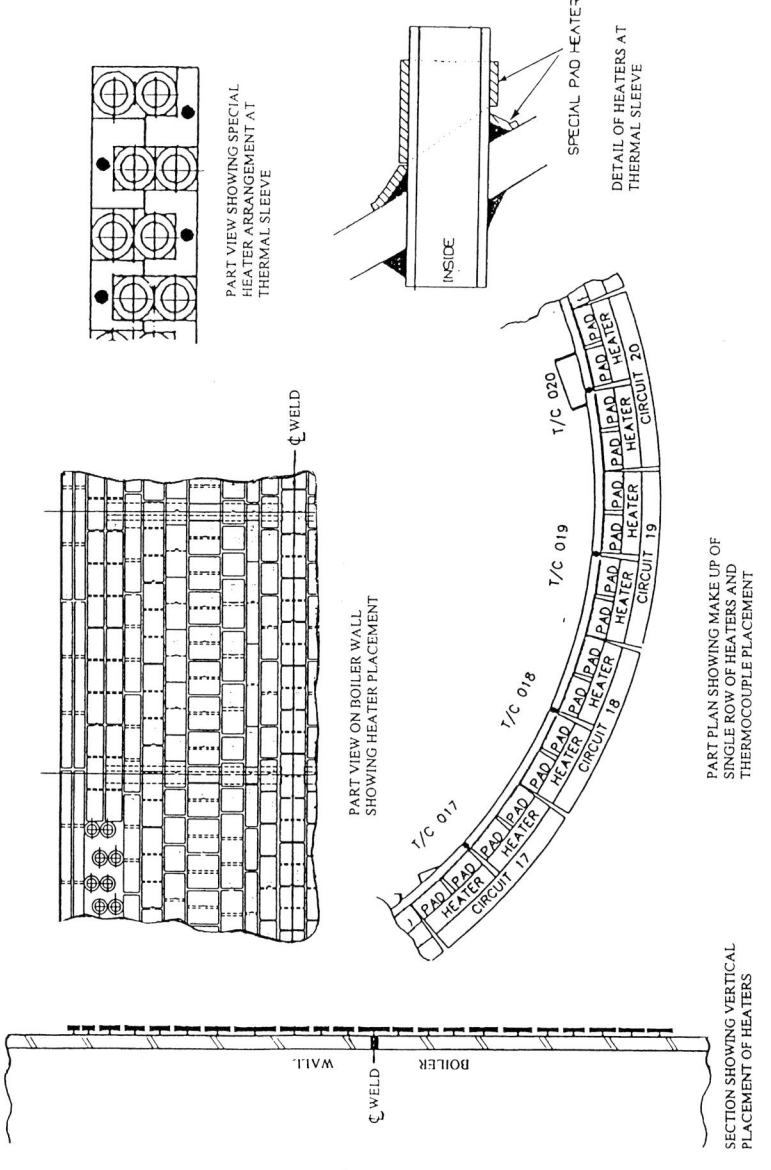

PART VIEW SHOWING SPECIAL HEATER ARRANGEMENT AT THERMAL SLEEVE

DETAIL OF HEATERS AT THERMAL SLEEVE

SPECIAL PAD HEATER

INSIDE

PART VIEW ON BOILER WALL SHOWING HEATER PLACEMENT

℄ WELD

T/C 017

T/C 018

T/C 019

T/C 020

PAD HEATER CIRCUIT 17

PAD HEATER CIRCUIT 18

PAD HEATER CIRCUIT 19

PAD HEATER CIRCUIT 20

PART PLAN SHOWING MAKE UP OF SINGLE ROW OF HEATERS AND THERMOCOUPLE PLACEMENT

BOILER WALL

℄ WELD

SECTION SHOWING VERTICAL PLACEMENT OF HEATERS

Figure 9: Part Plan showing make up of single row heaters and thermocouple placement

S690/008/99

Materials challenges

A N R HUNTER
Mitsui Babcock Engineering Limited, Glasgow, UK
M LAMB, E J McDONALD, and **R MOSKOVIC**
BNFL Magnox Generation, Berkeley, UK

ABSTRACT.

An extensive mechanical test programme has been carried out to determine the most appropriate PWHT temperature and to provide materials property data for safety assessments. The limited availability of material extracted from the plant meant that that a variety of similar materials and simulation techniques had to be employed and specimen sizes were limited. The measured data was consistent with expectations based on earlier work on Sizewell boiler materials and other similar materials. All of the results were validated by at least limited tests on representative material removed from the plant.

1.   INTRODUCTION.

Following the discovery of significant defects in three circumferential seam welds in the boilers at Sizewell Power Station it was decided that they would be subject to post weld heat-treated weld repairs (1). One aspect of the repair programme was to address a wide range of materials challenges. This included understanding of the metallurgical behaviour of the boiler materials and its significance in terms of the original cracking mechanisms, the repair process and the resulting mechanical properties of the boiler shell materials in the condition in which they were returned to service.

The boilers, Figs 1 and 2, were constructed from a low alloy Mn, Cr, Mo, V ferritic plate steel, with proprietary name BW87A with the composition given in Table 1. The main axial and circumferential welds, depending on location, were made using either manual metal arc (MMA) or submerged arc (SAW) welding processes. During original construction the welds were given at least one stress relief post weld heat treatment (PWHT) of three hours at 600°C. Some welds, or parts of welds, within sub-assemblies were stress relief heat treated during both initial assembly and then again during final site fabrication.

The investigation of the defects (2) had identified that they were due to stress relief cracking during the manufacturing PWHT. The cracking occurred because the creep ductility of the grain coarsened heat affected zone (GCHAZ) was too low to accommodate the strains arising from the combination of welding residual stresses and imposed PWHT thermal stresses. Lower creep ductility at what might otherwise be an ideal PWHT temperature is a characteristic of vanadium containing low alloy steels and is due to the formation of vanadium rich precipitates. Although it was proposed to eliminate GCHAZ from the repair weld and to minimise thermal stresses there remained a need to identify a PWHT which would produce higher ductility, to ensure that: (a) the repair weld has a minimum risk of stress relief cracking, and (b) original weld metal left after the excavation would have a low risk of stress relief cracking. Freedom from any form of defect initiation and growth in any part of the repair weld or adjacent original weld was of paramount importance to ensure that a weldment of the required high quality would be produced to meet the requirements of the safety case.

Although remote from the main defects the welds were generally of good quality they did contain small ultrasonically detectable defects. It was therefore necessary to derive acceptance criteria for any such defects in the ligament remaining after the weld repair excavation.

A safety assessment (3) for the repaired boilers will require mechanical properties for all materials (weld metal, HAZ, parent plate) in their various thermal history conditions: new weld metal, original material subjected to the repair PWHT temperature cycle, original material unaffected by the repair. Some of the data required for the return to service safety assessment was already in place in a register of approved materials data maintained within BNFL Magnox Generation (Magnox). However, additional data for the new weld metal and the subsequently derived PWHT condition had to be obtained from a dedicated test programme.

The purpose of this paper is to describe the materials test programme, which is summarised in Table 2, put in place to determine an acceptable PWHT regime and acquire materials properties data for the return to service safety case.

2   TEST MATERIALS.

2.1 Ex-plant and archive materials.
In many cases, determination of mechanical properties cannot be made on material taken from Magnox plant, and given its age, supplies of material put aside during construction are at best now extremely limited. Changes in steel making practices since the time of construction usually mean that equivalent materials are no longer commercially available. Hence it is necessary, where possible, to have special casts of material or weld consumables made to reproduce those used originally, albeit at considerable cost. In the case of the boiler shell repairs these constraints also applied, and during the early phase of the project no archive BW87A steel or weld metal were available. As a consequence use was made of a supply of Ducol W30 plate, which has a virtually identical composition to BW87A, but with a higher carbon content. The Ducol W30 had been removed from redundant sections of another plant. The advantage of this source was that it had operated at 360°C – 410°C and had thermally aged. Later in the programme, a ring of BW87A material was removed from the inlet cone of boiler 2D, the composition is given in Table 1. In addition, special cast plates reproducing the

original specification were made, primarily for the welding development (4) and they were also used for mechanical testing.

Reproduction welds were made using consumables specially manufactured to the contemporary specifications and deposited in accordance with the original welding parameters. When ex plant weld samples from: (a) within the weld preparation, and (b) butt welds in the ring of inlet cone became available, mechanical property tests were carried out to confirm that the data obtained from the reproduction welds were consistent with the plant materials.

Repair weld metal was obviously not a problem and plentiful supplies of material were available from the numerous weld procedure and welder training welds.

HAZ mechanical properties, particularly of the GCHAZ, are important for the structural integrity assessments, and accordingly had to be evaluated. Samples of HAZ from the plant did not become available until late in the repair programme (5) but in any event are particularly difficult to test because of the need to locate accurately the crack tip in the appropriate narrow region of the microstructure. In order to obtain baseline data, simulated HAZ material was prepared either by simple laboratory heat treatment or by producing a more appropriate thermal cycle using a Gleeble weld simulator (6). The derivation of process parameters for producing reproduction HAZ are described below.

It is known that prolonged service at temperatures approaching the maximum reactor gas outlet temperature (360°C) may result in thermal embrittlement, principally the migration of impurity elements such as phosphorous to prior austenite or ferrite grain boundaries resulting in an increase in brittle to ductile fracture transition temperature. In the absence of significant amounts of ex-plant material a step cooling heat treatment which reproduced thermal ageing was developed to investigate this parameter.

2.2 Scoop Samples from the Boiler Shells.

Boiler shell scoop samples (about 25mm diameter and 4mm thick) were extracted from regions of the boiler shells that had operated at high temperatures, which may have had grain boundary segregation, and from regions that had operated at sufficiently low temperatures to represent the start of life condition. Samples were removed from: parent plate, plate HAZ, manual metal-arc weld metal and submerged-arc weld metal.

The parent plates showed polygonal ferrite grains with colonies of tempered bainite. The plate HAZs showed tempered bainite or martensite in coarse grained regions and grain refined regions. The welds (both types) consisted predominantly of coarse columnar grains of acicular and proeutectoid ferrite with coarser grains showing in the submerged-arc weld metal. Auger electron spectroscopy (AES) analysis of the scoop samples removed from the boiler shells showed that in the high temperature regions phosphorus and molybdenum had migrated in service to the grain boundaries in all materials, see Table 3.

2.2.1 Hardness Testing.

For the parent plates and both kinds of weld, the average hardnesses of the scoops from high temperature and from the low temperature regions of the boiler shells were within the range of 5 HV_{10}, which was considered to be negligible. However, the average hardness of the HAZ in scoops from high temperature regions (297 HV_{10} based on six scoops) was higher than that for the low temperature regions (275 HV_{10} based on three scoops), which may point to an influence of thermal-ageing on the hardness (and yield stress) of the HAZ.

2.3 Simulated material.

Long term exposure of low alloy ferritic steels at temperatures in the range from approximately 340°C to approximately 570°C (7), promotes intergranular fracture which increases the ductile to brittle transition temperatures in fracture toughness and Charpy impact energy tests. Although upper-shelf fracture toughness and tensile properties are generally unaffected, other mechanical properties influenced by grain boundary strength, such as creep crack growth resistance, and near-threshold fatigue crack growth can be changed. This metallurgical effect is caused by the migration of minor alloy and impurity elements to grain boundaries. Known embrittling elements, in decreasing order of influence, are antimony, sulphur, phosphorus, tin and arsenic. The loss of fracture toughness may be eliminated by heat-treatment for about one hour at temperatures above 600°C (8). Thus tempering the boilers at 600°C would not have induced thermal embrittlement. However, cooling from the tempering temperature and long term exposure of the upper regions of the boilers to service temperatures of 340°C to 400°C may have resulted in segregation of these elements to grain boundaries. A programme was instigated to determine the extent of thermally-induced solute segregation in the boiler shells and to develop a heat-treatment to simulate material in the embrittled condition.

2.4 Simulation heat treatment.

2.4.1 Thermal ageing.

Although a source of thermally aged parent plate was available, a heat-treatment was needed to simulate the conditions at the end of life for the other boiler shell materials. Heat-treatment sequences have been devised based on models of segregation in which, at higher temperatures, the thermal diffusivity of embrittling elements, principally phosphorus, is high and the equilibrium solute concentration low, and vice versa (7). Therefore, low near-equilibrium grain boundary concentrations of solute develop in short times at high temperatures, and longer times are needed to raise grain boundary concentrations at lower temperatures. For this study, a finite difference calculation based on the Langmuir-McLean model of equilibrium segregation (9) was used to develop an embrittling heat-treatment:

i.	525°C for 24 hours
ii.	500°C for 144 hours and
iii.	475°C for 504 hours.

This heat-treatment and a de-embrittling heat-treatment (600°C for one hour and air cool) was applied to the thermally aged Ducol W30 parent . The results of AES undertaken on fracture surfaces to establish grain boundary compoition are given in Table 4.

The bulk phosphorus levels were higher in the boiler shells (at ≈0.02 at% in the plate and ≈0.03 at% in both weld metals) than in the simulation materials (≈0.016 at% in the plate and ≈0.03 at% in both weld metals). In the simulated HAZ, step-cooling heat-treatment induced levels of grain boundary phosphorus and molybdenum that exceeded those for the high temperature regions of the boiler shells (compare Tables 3 and 4). After step-cooling the manual metal-arc weld metal showed similar phosphorus and molybdenum levels to those in the high temperature regions of the plant, whereas, for the submerged-arc weld metal, the step-cooling heat-treatment reproduced the molybdenum level in the thermally aged boiler shell material, but induced a relatively low level of grain boundary phosphorus. Overall, on the basis of these observations the step-cooling heat-treatment was judged an adequate method of simulating the Sizewell A boiler shell materials in the service exposed condition.

3.   HAZ MATERIAL FOR TESTING.

Initially thermal simulations of HAZ microstructure were effected by subjecting material to an appropriate austenitising heat treatment in a laboratory furnace and then cooling them through the transformation temperature range between 800°C and 500°C at the same rate as experienced by the region of the HAZ of interest. The thermal cycles produced by heat treatment of large blocks generate only a minor limited range of microstructures and may not replicate accurately the microstructure of HAZ obtained during welding. Although these factors were found to have little effect on creep properties, they had a significant influence on fracture toughness behaviour. As an alternative test specimens were produced by thermal simulations employing a Gleeble 1500 (6) thermomechanical simulator.

3.1 Thermal simulations of heat affected zone.

HAZ microstructures were simulated by controlled heating and cooling in a Gleeble 1500 thermomechanical simulator (6). The conditions were based on thermal modelling, and comparisons between the hardnesses and microstructures of the simulations and the boiler shells. Three heat-treatments were applied to inlet cone material to allow for variations in the microstructure of grain-coarsened HAZ in the boiler shells:

   i. A predominantly martensitic microstructure was produced by heating BW87A plate to a peak temperature of 1350°C , and cooling from 800°C to 500°C in five seconds.

   ii. A mixed bainitic - martensitic microstructure was produced by heating BW87A plate to a peak temperature of 1350°C and cooling from 800°C to 500°C in fifteen seconds.

   iii. A predominantly bainitic microstructure (mixed with some martensite) was produced by heating to a peak temperature of 1350°C (as previously) and cooling from 800°C to 500°C in twenty-four seconds.

In multipass welds, the thermal cycle due to the deposition of subsequent weld beads heat treat part of the HAZ from the previous bead. To assess the effect of an additional thermal cycle on the properties of GCHAZ, two batches of specimens were subjected to an additional thermal simulation that involved heating the specimens to a peak temperature of either 1000°C or 775°C. For both peak temperatures, the cooling rate corresponding to cooling from 800°C to 500°C in 5 seconds was used for both the initial and the final thermal cycle.

3.2 Ex-plant weld HAZ.

Weld HAZ was extracted from plant and used to manufacture test specimens to provide verification of trends observed in fracture toughness data measured on thermal simulated HAZ. Due to the narrow width of the HAZ and the coarse grained region in particular, it was difficult to locate the notch and the associated fatigue crack of the fracture toughness specimens in GCHAZ and hence only a limited number of tests could be carried out. Tests on ex-plant weld HAZ were used to verify the trends in fracture toughness observed on thermally simulated HAZ due to test temperature and tempering treatments.

4.   DETERMINATION OF THE OPTIMUM PWHT TEMPERATURE

A specific test programme was undertaken to establish the heat treatment cycle for the PWHT. This required that both tensile and creep properties were established over a range of possible heat treatment temperatures. Subsequently, stress-relief heat-treatment of a mock-up weld was carried out to confirm the suitability of the heat-treatment.

4.1 Tensile Tests.

The tensile tests were carried out on specimens taken from the inlet cone. Blocks of approximate dimensions 30mm × 80mm × 90mm were extracted and heat-treated to simulate the grain coarsened heat-affected zone (GCHAZ) microstructures by austenitising at 1030°C for 30 minutes and quenching in warm water (50°C).

As expected, the tensile strength properties decrease over the temperature range 550°C to 700°C, see Fig.3. Although there is random scatter, the measured reductions in area and elongations, Fig 4, show a discernible trough at about 600°C and rising values up to 700°C, where the elongation was 61% and the reduction of area was 90%.

4.2 Creep Rupture Tests.

A series of creep rupture tests was carried out on simulated HAZ material similar to that used for the tensile tests, over the temperature range 550°C to 700°C at stresses which were approximately 0.75 and 0.85 of the (elevated temperature) 0.2% proof stress. The rupture strains (for both ratios of test stress to proof stress) are less than 3% at temperatures below 625°C and increase monotonically to over 10% over the temperature range from 625°C to 700°C.

Additionally, creep rupture tests were carried out on HAZ simulated on special casts of BW87A plate steel and heat treated as mentioned earlier. These were tested at temperatures in the range from 400°C to 700°C. Tests were carried out at 75% and 100% of the 0.2% proof stress appropriate to the test temperature. The behaviour was similar to that of the inlet cone material, with low ductility (less than 3%) in the tempering range from 400°C to 625°C. At higher temperatures, from 625°C to 700°C, there was a relatively rapid rise in ductility, attaining values greater than 10% and 30% at 625°C and 700°C respectively. Below a temperature of 625°C, the specimens failed with little evidence of necking and there was a tendency for failure to occur at stress concentrating features indicating a sensitivity to stress state.   In contrast, specimens tested at temperatures above 650°C exhibited significant necking. Scanning electron microscopy showed the fracture surfaces to be intergranular; at temperatures less than 625°C with smooth grain surfaces, but as temperatures approached 700°C, they showed evidence of microplasticity.

4.3 Creep Crack Growth Tests.

4.3.1 Specimens and procedure.
Creep crack growth tests were carried out on simulated HAZ, HAZ in reproduction welds and HAZ extracted from the boiler shells during the weld preparation.

Gleeble-Simulated HAZ

Single edge notch bend (SENB) specimens with 10mm square end faces were produced with predominantly martensitic and predominantly bainitic microstructures HAZ microstructures using the Gleeble simulator, as described earlier.

HAZ in BW87A Reproduction Plate

Material was removed from weld procedure development weldments (3), which was made in the special reproduction cast of BW87A. The weldment was post-weld heat-treated for a minimum of two and a half hours in the temperature range from 590°C to 610°C, to simulate the post-weld heat-treatment applied to the boiler shells during construction. It was then subjected to the step-cooling sequence described earlier to simulate thermal ageing in service. Specimens were either 10mm thick SENB or 12.7mm thick compact tension (CT) geometry.

HAZ Extracted from Boiler 2A, Sizewell A

In the course of undertaking the excavation for the repair, material was extracted from an undamaged HAZ. Two SENB specimens were produced, with the same geometry as those from the reproduction plate, above.

4.3.2 Test Procedures
Constant load creep crack growth tests were carried out on the Gleeble simulated HAZ and the reproduction weld HAZ at temperatures over the range 450°C to 700°C. The specimens from the boiler shell were tested late in the programme at the, by then, proposed repair post weld heat-treatment temperature of 675°C. Testing followed the relevant standard (ASTM E 1457-92). After each test, a cut was made parallel to the side faces of the specimen. The piece from one side of the cut was chilled in liquid nitrogen and broken open for fractographic examination and determination of the final crack extension. The remaining piece was prepared for metallography.

4.3.3 Results.
Gleeble simulated HAZ

At temperatures between 450°C to 650°C, the tests showed high creep crack growth velocities, similar to those for coarse-grained HAZ in 0.5%Cr-Mo-V steel at 565°C. There was a tendency to lower velocities at the extremes of this range. At 675°C and 700°C, the creep crack growth rates were lower than at the lower temperatures, approaching the rates for carbon-manganese steel plate. Fracture was intergranular, and showed increased branching at higher temperatures ≥650°C until blunting was observed at 700°C. At temperatures ≤ 500°C the crack tip opening displacements (CTOD) was ≈50μm, at 600°C, the CTOD was about 10μm and at 700°C the CTOD was approximately 100μm. Two HAZ microstructures, martensitic and bainitic were studied and showed similar creep crack growth behaviour.

Weld HAZ in reproduction plate.

For these tests, the pre-cracks were positioned mostly in the coarse-grained HAZ. The optimised repair weld exhibited superior ductility to initiation of cracking, during crack growth and at failure than the original weld, as indicated by its (higher) load line displacements. For both welds, ductilities increased significantly over the temperature range 625°C to 700°C. They supported a trend for which, at low temperatures, 400°C to 550°C, the crack growth rates were low due to the low creep rate, whilst at 675°C and above the crack growth rates were reduced as a consequence of the increased ductility.

Samples from the Defective Weld between Courses 6 and 7 in Boiler 2C

The creep crack growth rates decreased during both the tests on ex-plant HAZ, starting close to the (high) rates for Gleeble-simulated HAZ at 675°C and ending below the recommended rates for carbon-manganese plates as the cracks propagated from GCHAZ into either grain refined HAZ (GRHAZ) or weld metal. Polished sections across the specimens from these tests showed intergranular cracking in the HAZs, and within 500μm of the weld fusion boundaries. In one of these tests, the creep crack had propagated into the weld fusion boundary, and the low crack growth rates at the end of this test may represent the behaviour of the weld metal (rather than the HAZ).

5. MECHANICAL PROPERTIES FOR RETURN TO SERVICE SAFETY ASSESSMENT.

The safety case for future operation of the boilers will address, amongst other things, the safety margins under all operating conditions both for all known defects and for hypothetical defects. For the repaired boilers, it was recognised that the existing mechanical property data was incomplete both in terms of new materials (the weld metal) and thermal history (effects of higher PWHT temperature than employed during original construction). In this context, creep properties and fracture toughness are the most important parameters and this section describes the relevant parts of the test programme.

For creep properties, proposals were initially based on data for closely related low alloy steel, Ducol W30. The recommendations were judged to be conservative, principally by analogy with the behaviour of related low alloy steels (including ½Cr-Mo-V, 1Cr-½Mo, 2½Cr-1Mo and 9Cr-1Mo steels). However, to allow for the range of microstructures in the welds and their spatial distributions, a series of tests was carried out to substantiate the recommendations. Similarly for creep crack growth, recommendations based on the behaviour of simulated GCHAZ in another closely related low alloy steel (BS 1501-271) have been compared with results from a series of tests on specimens from the boiler shells or simulating the boiler shells.

With regard to the fracture toughness properties, they were measured, as far as possible, on samples cut out from the boilers. This approach ensured that the microstructure, inclusion content and the chemical composition of the steels, manufacturing processes and service history were representative of the boiler shells

5.1 Creep.

5.1.2 Creep Rupture.
Recommendations on the stress rupture properties of the boiler shells were based data from

numerous sources (10 - 14). An equation of the form:

$$\frac{\log_{10} t - F}{T - G} = a + b \log_{10} \sigma_r + c(\log_{10})^2 + d(\log_{10} \sigma_r)^3 + e(\log_{10} \sigma_r)^3 \qquad ...(1)$$

was chosen to represent the stress rupture behaviour, where: t is time in hours, T is temperature in °K and σ_r is rupture stress in MPa. a-e, F and G are numerical terms which describe material behaviour, their values were estimated by least-squares regression. The predictions gave an accurate or slightly pessimistic description of most of the test results, within the stress range 56MPa to 350MPa, for which they were recommended.

The recommendations for plate were extended to cover HAZ and the repair weld metal by assuming behaviour in line with that for other low alloy steels. For instance 1Cr-½Mo steel shows rupture stresses for weld metal that are about 10% greater than those for the steel in the normalised and tempered condition (15) and coarse-grained HAZ in ½Cr-Mo-V steel is more resistant to stress rupture than is the parent plate (15).

Subsequently mechanical tests have been performed on MMA weld metal, BW87A inlet cone steel and cross weld specimens from a simulated repair weld. These tests showed a reduction of creep rupture stress by about 50MPa that resulted from the high temperature PWHT, but otherwise supported the original recommendations as a conservative basis for assessments.

5.1.3 Creep Strain.
Creep strain test data were taken from the same studies that provided the basis for the creep rupture recommendations (10 – 14) and used to derive a predictive equation of the form:

$$\varepsilon = \varepsilon_t [1\text{-exp} (-\alpha \dot\varepsilon_s t)] + \dot\varepsilon_s t \qquad ...(2)$$

Here ε represents the strain resulting from primary and secondary creep: ε_t is the primary creep strain, which is taken to be the ratio of the applied stress to the yield stress, α is a constant, $\dot\varepsilon_s$ is the secondary creep strain rate (per hour) and t is time. Secondary creep strain rates were predicted from:

$$\dot\varepsilon_s = A[\sinh (\beta \sigma)] \exp[\frac{-Q}{RT}] \qquad ...(3)$$

in which σ is stress (MPa), Q the activation energy for self-diffusion of iron, R the gas constant and T temperature (K). A, B and C are constants determined by regression.

As with stress rupture, these predictions (based on plate properties) were extended to cover the HAZs and welds for behaviour in line with that of other low alloy steels. An indication of the creep strain behaviour of HAZs in the Sizewell A boiler shells is given by data for ½Cr-Mo-V steel at 565°C, for which parent material shows strain rates approximately six times those for simulated coarse-grained HAZ. Tests on 1Cr-½Mo steel and 2¼Cr-1Mo steel at 565°C, show strain rates in the weld metals that approximate to lower bound rates for the wrought materials and equations describing the creep strain of these materials in the wrought condition provide conservative overestimates of creep strains in the welds.

These recommendations are being underwritten by ongoing long-term tests on H1 manual metal arc weld metal and BW87A plate steel .

5.1.4 Creep Crack Growth.

Results were available from creep crack growth tests on simulated weld HAZ in BS1501 Grade 271 plate, which has a similar composition to the Sizewell A boiler shell plates (16). The material used in these tests was made from 52mm thick BS1501 Grade 271 plate in the normalised and tempered condition (with normalising at 880°C for 15 minutes, air cooling, tempering at 670°C for one hour, and air cooling). There were differences in composition between this simulated HAZ material and the Sizewell A boiler shell plates, which might have been significant:

i)   The average carbon level of the boiler shell plates (0.11%) was less than that of the simulated HAZ (0.14%).

ii) The average phosphorus and sulphur levels in the boiler shell plates (0.018% and 0.015% respectively) were higher than those of the test plate (0.008% and 0.002% respectively).

iii) The vanadium content of the boiler shell plates (0.11%) was higher than that for simulated HAZ (0.08%).

The simulated HAZ was produced by austenitising the plate at 1150°C for 30 minutes, water quenching and post weld heat-treating at 630°C for two hours. The side-grooved compact tension specimen geometry was adopted with a nominal specimen thickness of 12.5mm. The tests were performed according to standard procedures (17) at 320°C and 360°C, under constant load conditions giving reference stresses in the range 613 - 781MPa. Creep crack growth rates from these tests were plotted against the deformation rate-dependent steady state creep parameter C^*. These results were compared with data for related carbon, carbon-manganese and low alloy steels to derive predictive equations.

In summary these data show that:

- fine-grained HAZ was more resistant to creep crack growth than GCHAZ.

- thermal embrittlement affected steady state creep crack.

- creep crack growth rates in H1 MMA weld metal and BW87A parent plate were bounded by the rates for GCHAZ.

- the bulk quench procedure adequately simulated GCHAZ.

All these tests have confirmed the applicability of the original recommendations.

5.2 Fracture Toughness Properties.

5.2.1 Testing Procedure.

Practical restrictions on the size of the samples removed from the boilers, required tests to be carried out mostly on SENB precracked Charpy geometry specimens together with some compact tension geometry specimens. The precracked Charpy geometry specimens were initially notched to a depth of 3mm. The notch was extended by fatigue to give a nominal

a_0/W ratio of 0.5, where a_0 is the initial crack depth and W is the specimen width. The fracture toughness tests were performed according to ESIS P2-92 test procedure (18). Each specimen was enclosed in an environmental chamber and either heated or cooled as appropriate to a predetermined test temperature at which it was held for up to two hours once the temperature stabilised. The temperature was controlled to ±2°C. The specimens were tested in displacement control at a constant rate of increase of stress intensity factor in the elastic regime.

Tests were terminated either by failure of the specimens as a result of cleavage instability or by unloading the specimens after a certain amount of stable crack growth. The unloaded specimens remained unbroken at the end of the test and were heat tinted to mark the ductile crack growth. After cooling the specimens in liquid nitrogen and breaking them open, the extent of ductile crack growth was measured using either scanning electron or optical microscopy. Crack measurements were made at the specimen edges and at seven equally spaced points across the specimen thickness. Values of J were evaluated according to the CEGB procedure (19)

5.2.2 Results

5.2.2.1 Heat Affected Zone
The weld HAZ is typically only 2 or 3mm wide and contains the recognised ranges of different microstructures. The GCHAZ is only few grains wide ≈600μm. Due to the narrow width of the HAZ, and the coarse grained region in particular, it is not easy to measure the fracture toughness of this specific material. It was to overcome these difficulties that the simulated HAZ material described earlier was produced.

Fracture toughness and Charpy impact properties were measured on specimens obtained from different thermal cycles simulating the HAZ microstructures. Prior to carrying out the mechanical testing, all specimens were tempered for three hours at 600°C. The mechanical properties obtained for the GCHAZ produced cooling using a time of 5 seconds from 800°C to 500°C were found to give the most conservative fracture toughness properties. Fracture toughness properties at 350°C measured on plane sided precracked Charpy geometry specimens for cooling times of 5 and 24 seconds are illustrated in Fig. 7 for TL notch orientation. To assess the effect of notch orientation on fracture toughness of GCHAZ further tests were performed on side grooved and precracked Charpy specimens. Since the crack planes in the boiler shells were approximately perpendicular to the plate surface, it was considered appropriate to compare the fracture toughness properties of specimens for LT and TS notch orientations. In the ductile to brittle transition temperature region, the cleavage fracture toughness for LT notch orientation was found to be higher than for TS notch orientation as illustrated in Fig. 8.

Tempering temperature is another variable that needed to be considered in relation to the weld repair of the boiler shell. Previous work within Magnox has shown that as the tempering temperature increases the ductile to brittle transition temperature for the Charpy impact energy of Ducol W30 decreases. Further work was carried out to assess the upper shelf fracture toughness properties. Tempering treatments of 3 hours at six different tempering temperatures were used: 550°C , 580°C, 625°C, 650°C, 670°C and 690°C. The room temperature Vickers hardness (HV_{10}) measured on tempered specimens was found to decrease with increasing tempering temperature from 347HV at 550°C to 243HV at 690°C. To assess the fracture toughness behaviour on the upper shelf, side grooved precracked Charpy specimens were tested at 350°C. The results are displayed in Fig. 9. It could be shown by statistical analysis that both the initiation fracture toughness and crack growth resistance increase as the tempering temperature increases.

5.2.2.2 BW87A Plate Steel

Plane sided 22.5mm thick specimens manufactured from BW78A inlet cone plate steel were used to measure the fracture toughness. Most of the tests were performed on material in the as-received condition using specimens with TS and TL notch orientations. The upper shelf behaviour was achieved at 0°C and 30°C, respectively. The upper shelf fracture toughness was noticeably higher at 20 to 30°C than at higher temperatures. Statistical analysis of fracture toughness data obtained at 220°C and 350°C showed that within the scatter, the test temperature and notch orientation had no discernible effect. To assess the effect of weld repair on fracture toughness, further tests were carried out at 100°C, 220°C and 350°C test temperatures on specimens notched in TS orientation and tempered at 675°C for 4 hours. These tests yielded initiation fracture toughness values approximately 30% higher than those for the as received condition.

5.2.2.3 Manual Metal Arc Weld

Fracture toughness properties of MMA weld metal are based on data measured on reproduction welds on 19mm thick plane sided compact tension specimens. Since the original test data are associated with tempering treatment of 3 hours at 600°C, additional data were obtained for tempering treatment of 3 hours at 675°C. These tests were performed on 25mm thick plane sided compact tension specimens machined from a reproduction MMA weld. The fracture toughness values of this weld metal at two different tempering temperatures are compared in Table 5. For both heat treatments, specimens were notched with the notch direction perpendicular to the free surfaces of the weld and the crack plane parallel to the weld length.

The temperature dependence of fracture toughness for the two tempering treatments is different resulting in the values of initiation fracture toughness which are higher for the 675°C than for the 600°C tempering temperature.

To compare the fracture toughness properties of the reproduction and the boiler shell welds, small coupons of MMA weld metal were extracted from boiler shells and reconstituted by electron beam welding into Charpy size specimens. The notch orientation was consistent with that used for the reproduction welds. Test specimens were side grooved to a total of 20% of the original thickness. Fracture toughness tests were performed at 100°C, 220°C and 350°C for both as received and heat treated conditions. The initiation values of fracture toughness fell generally below those obtained for the reproduction welds. One reason for the disparity is the small specimen size used for materials taken from the boiler shells.

The microstructure of multipass welds is heterogeneous; the two main microstructural features in a single weld bead are the as cast columnar grains and the fine-grained reheated region, Fig. 10. Because the ligament of a Charpy specimen is comparable to the size of a weld bead some specimens are likely to sample a single microstructural region and give rise to extreme values of fracture toughness. Metallographic examination of broken precracked Charpy specimens showed that the proportion of coarse grained microstructure across the specimen thickness varied between approximately 50 and 85%. It was found that for specimens in the as received condition tested at 220°C and 350°C it was possible to correlate the magnitude of initiation fracture toughness with the proportion of coarse grained columnar grained microstructure intersected by the crack tip. The largest coupons of MMA weld metal were reconstituted into 19mm thick compact tension specimens and used for testing at 350°C. The fracture toughness values measured on these specimens were in the range from 88 to 150N/mm. Comparison of initiation fracture toughness values measured on ex-plant precracked Charpy specimens showed that tempering at 675°C brought about a significant increase in initiation fracture toughness.

6.   DISCUSSION.

An extensive materials property evaluation programme has been carried out with the primary objectives of determining an appropriate PWHT regime and to provide mechanical property data to be used in safety assessments of the plant when it is returned to service.

Availability of material meant that for the majority of the work some compromises had to be made on material source or specification and specimen dimensions. Techniques for providing substantial amounts of material which would otherwise be in limited supply, or very difficult to repeatably test, e.g. HAZ were developed and the validity of the ensuing test data demonstrated. Thermal aging during service is an important parameter for some properties and in order to increase the amount of suitable material for testing purposes a short term heat treatment cycle which induced levels of aging comparable to those attained during 30 years operation was developed and validated. Where data were obtained on other than ex-plant material they were validated by comparisons with known behaviour and behaviour trends in similar materials and confirmatory tests on material removed from the plant.

In the case of the selection of a PWHT regime the various tests carried out in the potential PWHT temperature range gave similar results, i.e. the GCHAZ possesses poor ductility and high creep crack growth rates at the original PWHT temperature (600°C), which was useful substantiation of the metallurgical analysis of the major defects. Properties improved with increasing temperature, but significant improvement did not start to occur until 625°C, thereafter the rate of improvement increased with temperature. In determining the ideal PWHT temperature there was no absolute criteria, and the basis of the selection was largely judgmental. Clearly temperatures below 625°C represented an unacceptable risk, as the properties would not be significantly better than during the original construction PWHT. Excessively high temperatures, on the other hand, were not automatically beneficial both in terms of ambient or operating temperature mechanical properties or the feasibility of the PWHT operation. Another part of the mechanical test programme, not reported in this paper, showed that while most mechanical properties improved with increasing PWHT temperature, tensile strength reduced with increasing PWHT temperature.

Although properties started to show improvement at about 625°C the change was initially gradual with PWHT temperature and it was not until temperatures in excess of 650°C were attained did the changes become significant and, in some cases alternate mechanisms become evident, e.g. crack tip blunting in creep crack growth tests and changes in fracture micromechanisms. Combining these experimental results with an allowance for variations due to composition and the feasible tolerance on temperature during PWHT resulted in a proposed PWHT temperature of 675 ± 10°C.

Materials properties for safety assessments during future operation mainly considered the new materials introduced and the changes due to the higher PWHT temperature. The programme was comprehensive and covered all possible combinations of material and thermal history. In general it was found that toughness improved with increasing PWHT temperature, and although there was a reduction in creep strength with increasing PWHT temperature it was acceptable. Creep rupture and creep ductility tests are taking place over a longer term than the repair programme, but initial results indicate that behaviour is consistent with the predictions based on the behaviour of similar materials.

7.   ACKNOWLEDGMENT.

This paper is published with the permission of the Director Technology and Central Engineering, BNFL Magnox Generation.

The assistance of R F Smith, M Priest and I J Lingham who carried out the bulk of the experimental work is acknowledged.

8.   REFERENCES.

1. Smitton C. and Marchese C. J. Paper 1 these proceedings.

2. Exworthy L. F, Flewitt P. E. J. and Ellis B. J. C. Paper 2, these proceedings.

3. Jeans P. J, Taylor E. G. and Munro H. G. Paper 5, these proceedings.

4. McDonald E. J, Hunter A.N.R and Bell W. M. Paper 6, these proceedings.

5. Evans H. V, McDonald E. J. and Wilkens A. W. Paper 9, these proceedings.

6. Gleeble Dynamic systems inc

7. Viswanathan R: "Damage mechanisms and life assessment of high temperature components," 1989, ASM International.

8. Viswanathan R: Metallurgical Transactions, Vol. 2, pp. 809-815, March 1991, "Temper embrittlement in a Ni-Cr steel containing phosphorus as an impurity."

9. McLean D: "Grain boundaries in metals", Oxford, Clrendon Press, 1957.

10. Lessels J and Barr R R: ISI Publication 97, 1967, Proc. Joint Conf. Organized by BISRA and ISI, Eastbourne, 4-7 April 1966, High temperature properties of steels, pp. 333-343, "The high temperature properties of a Mn-Cr-Mo normalised and tempered steel, Ducol W30."

11. Plastow CEGB Report RD/L/R991, 20 February 1961, "Creep and stress rupture properties of Colvilles' Ducol W30 Steel as three inch thick plate."

12. Willliams S R: 1960: CEGB Report RD/P/N4, "Review of the properties of Ducol-type steel in view of its possible use for reactor pressure vessels."

13. Hacon J CEGB CERL Laboratory Report No. 858, 13 April 1959, "Creep and stress rupture properties of Colvilles' Ducol W30 steel as 1½-inch thick plate."

14. Glen and Hazra L K: The Presentation of Creep Strain Data, 5-6 October 1971, pp. 61-94, 1972, Proc. Symp. Organized by British Steel Makers Creep Committee, "Some information on the creep behaviour of low alloy steels."

15. Cane, B J: CEGB Report RD/L/2101N81, August 1981, "Collaborative programme on the correlation of test data for high temperature design of welded steam pipes: presentation and analysis of materials data."

16. Maskell, R V: Mitsui Babcock Report 65/96/43A, 21 August 1995, "Creep Crack Growth in Carbon Manganese Steels at 300 C -420 C."

17. American Society for Testing and Materials, Subcommittee E24 on Subcritical Crack Growth: ASTM Standard Test Method E1457-92, April 1992, pp. 1031-1040, "Standard Test Method for Measurement of Creep Crack Growth in Metals."

18. ESIS P2- 92

19. Neal B K, Curry D A, Green G, Haigh J R and Akhurst K N: A procedure for the determination of the fracture resistance of ductile steels, Int. J. of Pressure Vessels and Piping, Vol. 20, No. 3, pp 155 - 179, 1985.

Table 1: Composition of the materials tested and of the Sizewell A boiler shell plates. (1) and (2) denote free and total nitrogen contents respectively.

Element	Content / Weight %					
	Specimens			Sizewell A boiler shell plates		
	Boiler 2D inlet cone (Cast BA83)	Test piece TP1 (Cast 3085)	Boiler 2A, Plate 6A (Cast BA112)	Mean from case histories	Specified limits for BW87A	
					Minimum	Maximum
C	0.10	0.12	0.11	0.11	0.100	0.15
Mn	1.30	1.51	1.36	1.33	1.20	1.50
Si	0.31	0.29	0.24	0.25	-	0.30
S	0.013	0.020	0.012	0.015	-	0.050
P	0.011	0.022	0.020	0.018	-	0.050
Cr	0.57	0.62	0.62	0.55	0.45	0.70
Mo	0.31	0.30	0.28	0.28	0.25	0.30
V	0.09	0.13	0.12	0.11	0.07	0.12
Ni	0.23	0.21	0.25	0.21	-	-
Cu	0.11	0.02	-	-	-	-
Ti	<0.002	-	-	-	-	-
Nb	<0.002	-	-	-	-	-
As	0.027	0.050	-	-	-	-
Sn	0.012	0.04	-	-	-	-
Sb	0.015	0.03	-	-	-	-
N	0.0122[1]	0.017[2]	-	-	-	-
O	0.0031	0.019	-	-	-	-

Table 2: Summary of Mechanical Testing Programme

Materials			Tensile	Fracture Toughness	Creep Rupture	Creep Strain	Creep Crack Growth	Charpy Impact
Plate	Ducol W30	SOL[1]	✓	✓				✓
		Aged[2]	✓	✓				✓
	Ex Plant BW87A	Aged	✓	✓		✓	✓	
		675 °C PWHT	✓	✓				
	Reproduction BW87A	SOL	✓		✓			✓
		Aged						
		675 °C PWHT	✓		✓			
Weld (ex boiler)	Ex Plant MMA	Aged	✓	✓				
		675 °C PWHT	✓	✓				
	Reproduction MMA	SOL	✓		✓	✓	✓	✓
		Aged						✓
		675 °C PWHT	✓	✓	✓			
	Ex Plant SAW	Aged	✓					
	Reproduction SAW	SOL						✓
		Aged						✓
		675 °C PWHT	✓	✓				
HAZ[3]	in Ducol W30	SOL						✓
		Aged		✓				✓
	in BW87A	SOL				✓	✓	
		Aged	✓	✓			✓	
		675 °C PWHT	✓	✓	✓		✓	
	Simulated in Ducol W30	SOL					✓	
		Aged	✓	✓			✓	
		675 °C PWHT	✓					
	Simulated in Ducol W30	Aged	✓	✓		✓		
		675 °C PWHT	✓	✓			✓	

Legend:
SOL[1] - start of life, either heat treated to de-embrittle or new unused reproduction material
aged[2] - either service aged or heat treated to simulate 30 years service
HAZ[3] - simulated - thermally simulated (gleeble) and bulk quench

Table 3: Grain boundary phosphorus, molybdenum and carbon contents from the Auger analyses of scoop samples removed from Sizewell A boiler shells.

Material	Thermal history	Composition, at. %		
		P	Mo	C
BW87A parent plate	Ex-service, T2	4.2	1.4	8.3
	Ex-service, below 300°C	0.3	0.4	4.0
BW87A plate HAZ	Ex-service, T2	4.7	1.6	6.3
	Ex-service, below 300°C	1.5	0.7	5.4
H1, manual metal-arc weld metal	Ex-service, T2	4.1	1.8	4.1
	Ex-service, below 300°C	2.0	0.9	4.9
Union Melt No. 4B, submerged-arc weld metal	Ex-service, T2	4.3	1.5	5.5
	Ex-service, below 300°C	1.8	1.0	4.4

Table 4: Auger analysis results on samples simulating the Sizewell A boiler shell materials.

Material	Source of sample	Condition	Composition, at. %		
			P	Mo	C
Ducol W30 parent plate	Dungeness A bellows unit C4B7	Ex-service, T2	6.5	1.6	4.4
		Ex-service, T2 and de-embrittled	2.0	0.8	6.7
Ducol W30 plate HAZ	Dungeness A bellows unit C2B9	Step-cooled	6.3	1.8	5.6
Babcock and Wilcox H1 manual metal-arc weld metal	Weld simulation	Step-cooled	4.2	1.5	4.0
Union Melt No. 4B submerged-arc weld metal	Weld simulation	Step-cooled	3.0	1.5	5.8

Table 5: The fracture toughness values of reproduction MMA weld metal for two different tempering temperatures

Test Temperature (°C)	$J_{0.2}$ (N/mm)		$K_{0.2}$ (MPa m)		Tempering Treatment
	Lower Bound	Mean	Lower Bound	Mean	
50-360	130	188	169	203	3 hours/600 °C
100	217	249	221	236	3 hours/675 °C
200	138	171	173	192	3 hours/675 °C
350	189	222	199	216	3 hours/675 °C

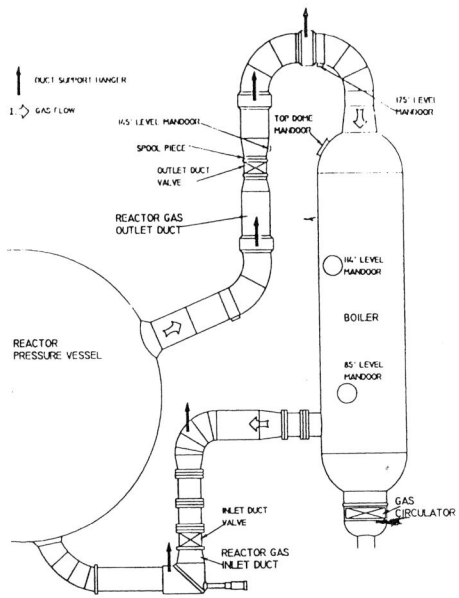

Figure 1: Schematic layout of gas circuits at Sizewell A

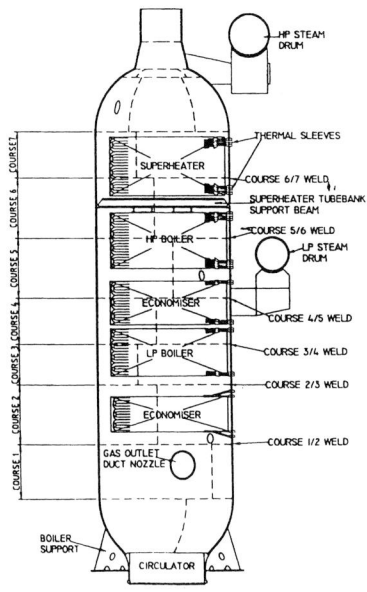

Figure 2: Schematic layout of boiler at Sizewell A

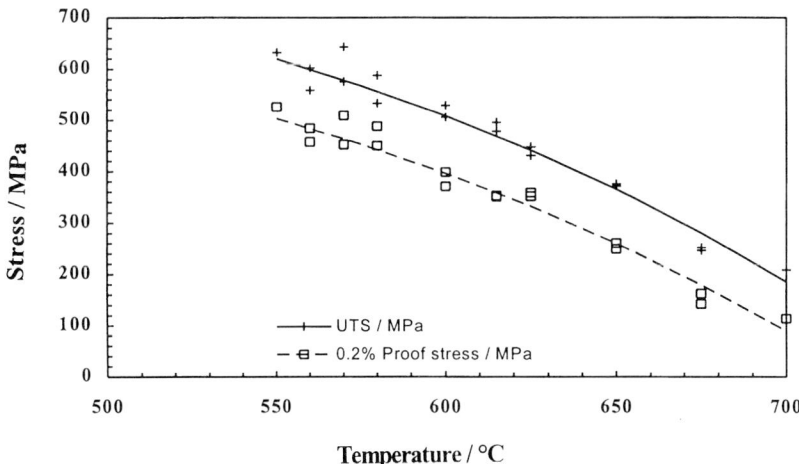

Figure 3   The temperature dependence of ultimate tensile strength (UTS) and 0.2% stress of simulated HAZ in BW87A steel plate.

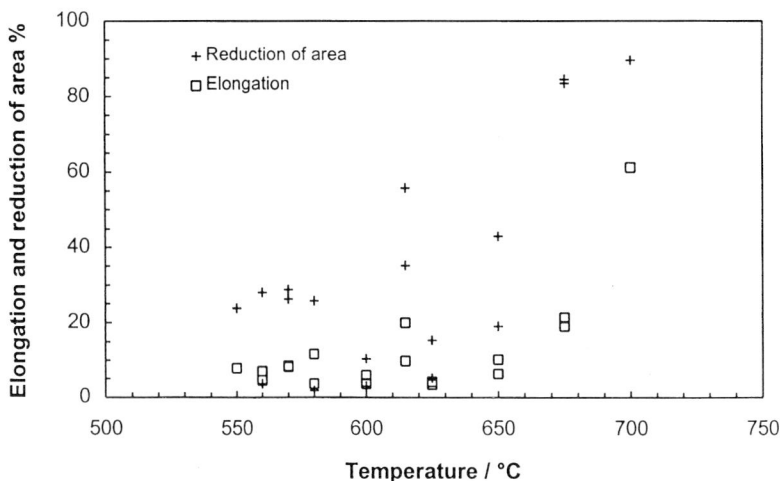

Figure 4   The temperature dependence of tensile ductility (as reduction in area and elongation) in simulated HAZ in BW87A steel plate.

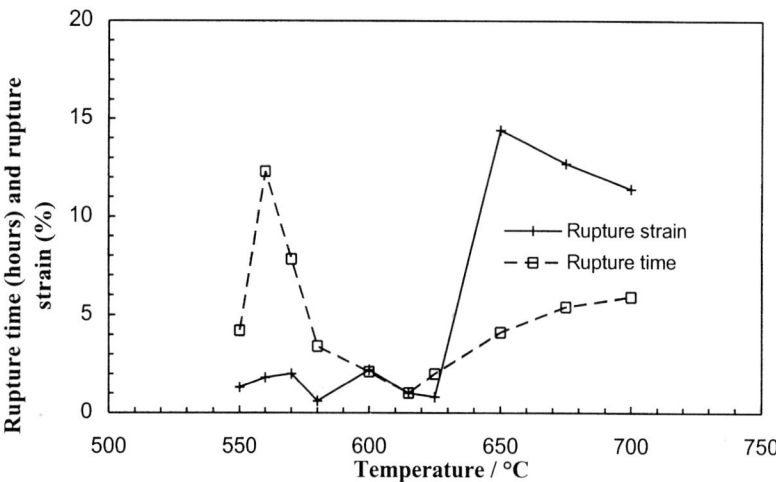

Figure 5 The temperature dependance of creep rupture strength and time for tests at aproximately 0.75 of the (elevated temperature) 0.2% proof stress.

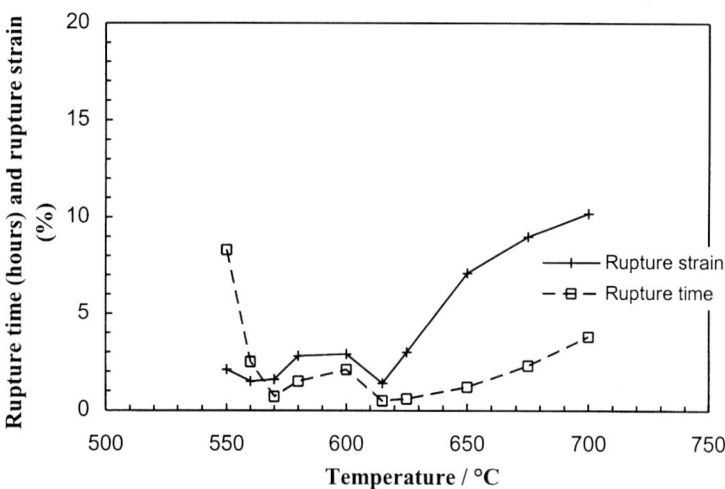

Figure 6 The temperature dependance of creep rupture strength and time for tests at aproximately 0.85 of the (elevated temperature) 0.2% proof stress.

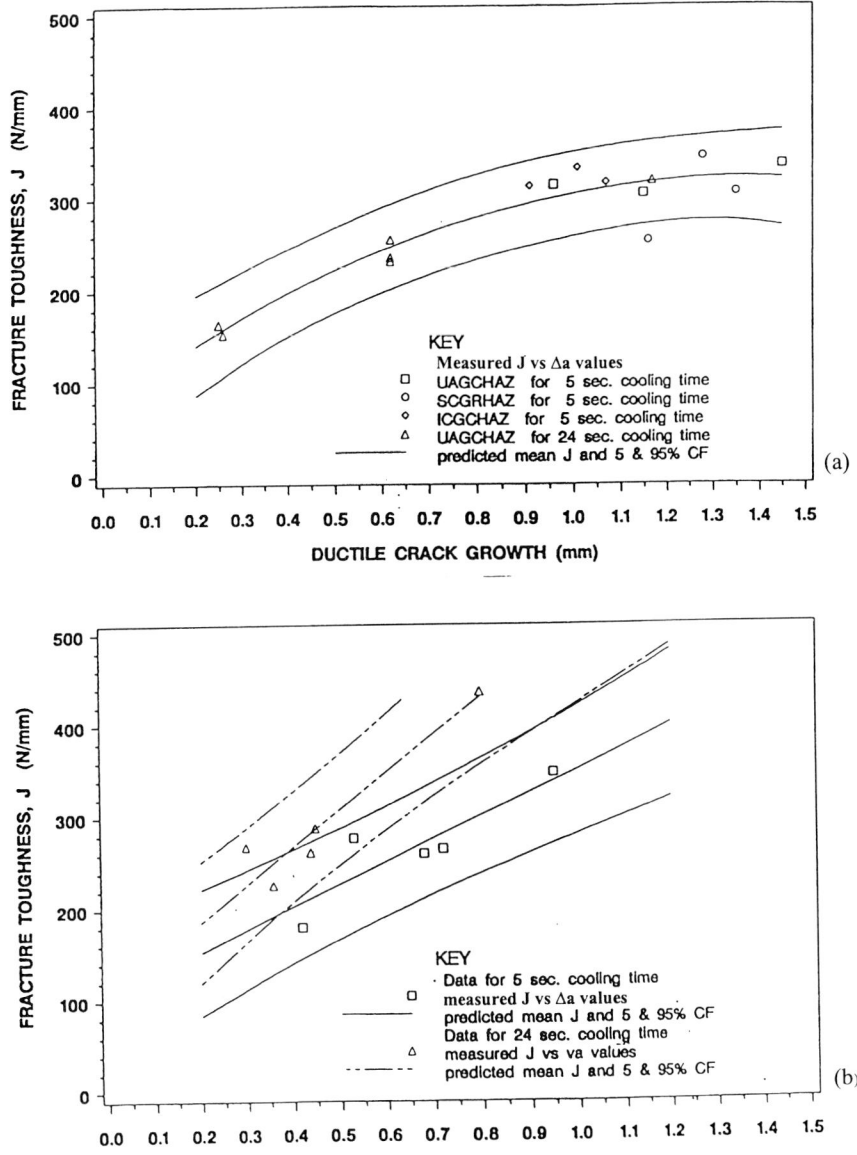

Figure 7 Measured and predicted J vs Δa crack growth resistance data for thermally simulated HAZ in (a) Ducol W30 tongue plate and (b) BW87A inlet cone material.

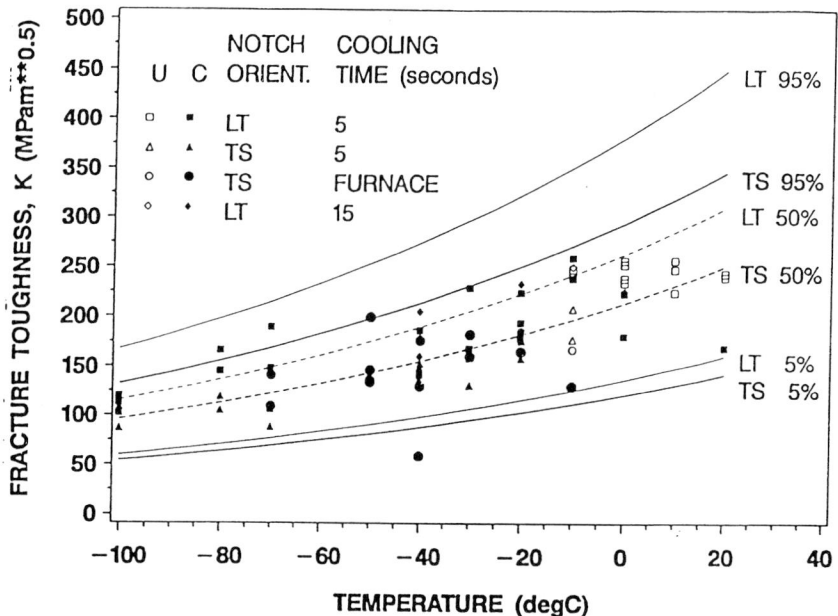

Figure 8   Comparison of 0.05, 0.50 and 0.95 quantiles computed for the joint distribution of K_c and Δa_c with the measured data.

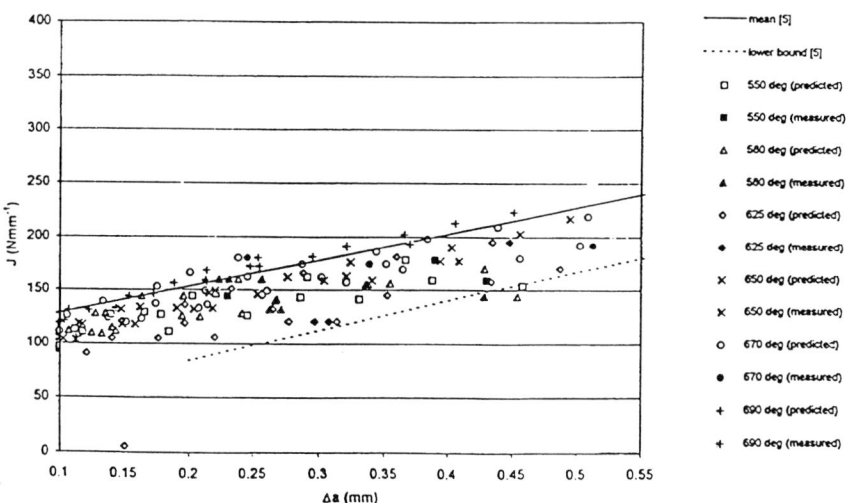

Figure 9 Shows the end of test J vs Δa pairs for each test, generated by the unloading compliance technique, for tempering treatment of three hours using six different tempering temperatures in the range from 550°C to 690°C. The mean and lower bound estimate of crack growth resistance are also shown.

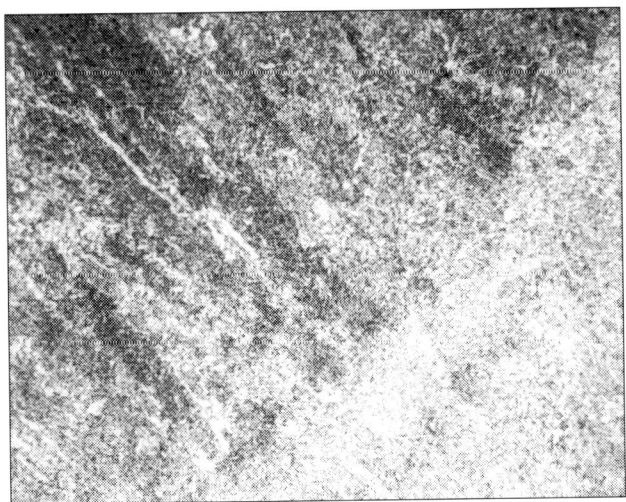

Figure 10 Typical microstructure of a single weld bead showing cast columar grained region and fine grained reheated region.

S690/009/99

Site project management and implementation of the boiler repair

H V EVANS and **E J McDONALD**
BNFL Magnox Generation, Leiston, UK
A W WILKENS
Mitsui Babcock Energy Services Limited, Gravesend, UK

ABSTRACT

This paper describes the key management control systems that were applied to the Boiler Shell Repair Project at Sizewell A and the methodology used to resolve technical challenges encountered during site implementation of the works. Project specific procedures were developed and were in addition to the normal control procedures of both the Site Licence holder, BNFL Magnox Generation (Magnox) and to the main contractor, Mitsui Babcock Energy Services Ltd (MBESL). The project was extremely complex with a changing work scope and a large number of different activities running in parallel, hence it was necessary to introduce very formal project management controls to ensure that safety and high standards of quality were maintained and a successful conclusion was reached. The paper also describes the more significant site implementation activities including the welding and post weld heat treatment operations.

1.   SITE PROJECT MANAGEMENT

In order to establish project control systems objectives of the project were defined and a full and independent review of the existing systems of Magnox and all of the main contractors was carried out. Using this approach it was possible to identify the strengths and more importantly the weaknesses of the existing systems to manage the particular requirements of the Boiler Repair Project. On completion of the reviews a set of key project management controls were introduced that integrated the systems of both Magnox and MBESL.

1.1 Project Objectives
1.1.1 Safety
"To undertake the works in a safe manner with no lost time accidents, in compliance with all statutory legislation and Local Site Instructions".

This was achieved by utilising the Construction Design Management (CDM) regulations and introducing specific safety management procedures and systems at the outset. All personnel requiring access to the project work areas for what ever reason were trained by a formal induction course and made aware of the work being undertaken and the hazards that were likely to be encountered to ensure that only authorised personnel entered the work areas. A site area pass system was introduced. Warning boards were placed at all entrances to the work areas detailing current activities and any specific new hazards. Tighter controls were introduced in local areas at specific times such as during welding or post weld heat treatment (PWHT) with guardians in attendance and only personnel authorised with a specific role were permitted entry.

A risk assessment of all of the project activities was produced and the relevant points were conveyed to the personnel by team talks prior to undertaking the activity. No work was permitted within the project area without the working party being in possession of the relevant risk assessment.

There was a project safety officer on site continuously throughout the repair programme to give advice on any safety issue and to ensure that the work areas were maintained in a clean and tidy environment.

During the risk assessments the project considered that fire was a major risk and a number of controls were put into place. The project employed dedicated Fire Wardens who were used to identify fire related hazards at the earliest opportunity. Strict controls on the use of potentially flammable materials, particularly in confined spaces were enforced. Additional controls, fire watchers and training were invoked during high risk phases, e.g. welding and PWHT.

To avoid electrical fires, a team of specialists monitored all electrical connections twice each day and continuously during PWHT, using Thermal Imaging cameras.

A further aspect of Safety were the Radiological controls relating to access into the boilers. Further, all debris and items removed were controlled and disposed via an approved Radiological disposal route.

All parties undertook regular site safety walks and daily toolbox talks were given to all personnel working in the areas. Safety culture was actively encouraged by management during site inspections and discussions of safety issues with the personnel at the place of work.

Other initiatives were the use of safety competitions and safety awards to maintain and stimulate interest in safety at the workface.

All of the above measures promoted a safety culture that was instilled throughout the whole project workforce which resulted in the achievement of the prime project objective. The project received a Gold ROSPA award on project completion for no lost time accident and demonstration of good safety management.

1.1.2 Quality

"To under take the repairs to a high quality with no reworking and to demonstrate to the Station Manager, Nuclear Installations Inspectorate and the Health Safety and Environmental Department that adequate control, had been exercised on the work".

During the early stages of the project a full site audit was carried out by the NII to ensure that a systematic approach to quality was in place to ensure that the stringent technical procedures for the repair could be carried out to the specified standard and recorded as such.

A Quality Steering Group was set up to oversee Quality arrangements whose membership included Magnox, Contractors and independent third party inspection agency (ITPIA) representatives.

One of the main objectives for the Steering Group was to ensure that there was a consistent and auditable interface between Station and Project quality arrangements to guarantee that existing nuclear site license requirements were maintained particularly with Reactor One plant which was operating at power throughout the duration of the project.

In order to control the quality of the work and to ensure that it was undertaken to the highest standards each operation was detailed in a method statement with a supporting quality plan to detail all of the inspection and witness points. Each method statement met the safety requirements of the corresponding stage submission. The quality plans and method statements were generated on site by MBESL and checked and approved by Magnox. The quality plans and method statements were reviewed by the Independent Third Party Inspection Agency (ITPIA) [Kennedy & Donkin Ltd.] who added their own inspection points. The documents were then reviewed by the Stage Submission Author to ensure that all relevant points had been addressed and final acceptance of the completed documents was by the Magnox site team.

All personnel involved with the site implementation were vetted to ensure that they were "Suitably Qualified and Experienced Persons" (SQEP) to carry out the work. A register of all personnel including workforce operatives was maintained and subject to Audit by the NII.

All operatives working on the plant had to be in possession of a work pack containing.

- Work Control Card.

- Method Statement.

- Quality Plan.

- Risk Assessment.

- Permit to work.

A central register was maintained for Technical Queries and Non Conformances and there was a formal procedure to ensure that these were all adequately addressed and closed out. They were tracked on a daily basis and were a standard agenda item at the daily site meeting.

Particular emphasis was placed on the case history for the project, including the site work, from an early stage to ensure an auditable trail of work done was available at all times. Project resources were allocated to case history preparation at the beginning of the work, this ensured that as each area of work was completed, full supporting documentation was also available. The approach resulted in the completion and acceptance of the case history within 7 days of project completion.

1.2 Programme

Prior to Magnox Board consent, a site activities programme had been put together by MBESL to reflect the durations of activities associated with the requirements identified in the Paper of Principle (1) which set out the principles of the repair operation and the work which would be carried out to assess and monitor the plant in order to be able to prepare a safety case for the return to power.

Site work started against this programme very soon after acceptance of the paper of principle and Magnox Board authorisation against this programme. Clearly at this stage, while there was confidence that the necessary activities would be carried out, the detailed workscope still had to be developed in some areas and the links with the technical development programme confirmed.

The project programme therefore comprised a high level component covering the whole project from which could be extracted detailed components covering activities to be undertaken in the short term. As the project moved forward there was a rolling expansion of the short term programme.

The project programme was used to monitor progress at the daily site meetings and the roll forward of the short term component was managed by the site technical team which comprised representatives of Magnox (Station staff, project site staff and Technology and Central Engineering staff), the main contractors and ITPIA. There was a formal process for the issue of revisions to the programme.

1.3 Documentation Control
The most fundamental system that was introduced for the project was control, flow and distribution of documentation. To manage this process a document control centre was set up. The quantity and variety of documentation that was produced to specify, control and record the work and also to support the safety case was significant. The documents were being produced in a number of different locations and all had to be tracked and controlled from a central location. A failure of this system would have seriously compromised the integrity of the repair project.

Each document was identified on the project programme and linked to site activities and registered in the project document control centre. Inevitably documentation was revised throughout the works and a rigid control system was introduced. The fundamental part of that control was that all changes should be highlighted and easy to identify. The latest revision of all documentation was recorded in the document control centre and a weekly schedule was produced so that all parties were aware of what had changed.

1.4 Division of work packages
Early in the project it was recognised that the site works could be halted while the relevant stage submission was being prepared and approved. This was not desirable as the project programme and costs would increase and more importantly the momentum of the site work would be lost. In order to overcome this problem a two-stage approach was undertaken.

Stage one was to view all of the prospective work under the Magnox guidelines to identify any elements that were non-nuclear significant and thus did not require to be approved in advance in a stage submission (but all of this work was formally reviewed prior to return to service). Operations such as the removal of bolted components and the preparation and initial inspection of non-repaired welds came into this category. This enabled the site works to proceed rapidly and to be proceeding in parallel with the preparation of the stage submissions and safety cases.

Stage two was to review all of the stage submissions and to divide them down into smaller units to permit a faster clearance of each item. A good example of this is the de-loading of the superheater tube banks. This was divided into three sections being manufacture, installation and the de-loading process. This approach whilst permitting a more consistent workload on site did have the down side of increasing the number of stage submissions required and introducing further complexities into the programme.

1.5 Site quality implementation

Quality plans and method statements alone do not guarantee that the finished work will be to the required standard. A full auditing and surveillance regime was introduced, with the work being monitored by MBESL supervisors and engineers, Magnox personnel, the ITPIA and the Nuclear Installations Inspectorate who made frequent site inspection visits and audits. Generally, surveillance and audit reports were favourable.

1.6 Communication

The most important aspect of managing a project of this nature is good communication and team working at all levels. An effective communication system was set up so that key personnel could be contacted on a 24 hour, 7 day week basis so that decisions could be taken to minimise any delays. The communication was formalised by the scheduling a number of meetings at different levels. At the basic level was a daily meeting, which reviewed all aspects of the site work. This was attended by all parties including support services. This encompassed all of the site works including safety and quality. Good communication was maintained with the work face by the use of radio pagers and hand held radios. At a higher level, a project Lead Team ensured that communications and setting of the project principles was managed efficiently. This was the prime decision making body.

2. SITE IMPLEMENTATION

2.1 Description of plant

The Sizewell A boilers are manufactured from a low alloy Mn Cr Mo V ferritic steel, BW87A, 57mm thick, 29.5m high and 6.9m in diameter and are connected to the reactor pressure vessel by 2m diameter mild steel ducts, Figs 1 and 2. There are 4 boilers arranged radially around each reactor. There are two boilers located in each boiler house.

2.2 Workscope

As stated previously the initial high level site workscope had been developed in parallel with the preparation of the paper of principle. Apart from the obvious elements such as the excavation, welding and heat treatment there were numerous supporting activities that needed to be undertaken. Many of these activities were associated with the heat treatment and its effect on the surrounding plant.

Using a simplistic approach of surveying the plant and utilising engineering judgement a list of plant components that had to be addressed was prepared. The following criteria were used for each item:

- Does it need to be modified to allow the weld repair or PWHT to be undertaken?

- Could the heat treatment process affect it and how?

- What are the counter measures?

- How can it be demonstrated that the plant has not been affected detrimentally by the repair process?

2.3 Changing workscope

By far the largest affect on both the programme and costs were the changes that occurred to the work scope due to the technical development work that was being undertaken. Some significant items were removed from the work scope and some equally significant ones added. All of the items in the original scope were challenged, as were all new additions. This resulted in the removal from the original scope of the requirement to de-load the high pressure steam drum from the boiler shell and also ensured that no new items were added that were not considered essential for either plant integrity or to support the safety case.

There were still some significant additions to the work scope that could not be avoided and which had a major impact on both the programme time and the cost. An example of this being the requirement to remove all of the 12 by-pass tubes from within each boiler instead of only 3 which were in the original scope. Further examples are the increase in the total length of the repairs from 8 to 66m and the significant increase in heat treatment extent and equipment required to minimise thermal gradients and maximise heating rates.

3. WORKSCOPE CONTENT

The scope and the reasoning behind the major items are detailed below.

3.1 External Mechanical Work

3.1.1 Preparation and inspection of shell welds

In order to demonstrate that the heat treatment process had not adversely affected the existing boiler shell and attachment welds, all welds that were subjected to a temperature greater than the normal maximum operating temperature of the plant were to be prepared and inspected, using ultrasonic and magnetic particle techniques, after the heat treatment cycle. In order to provide a baseline for the post PWHT inspection all these welds were subject to inspection prior to PWHT. This included the superheater inlet pipework thermal sleeves and all accessible internal and external attachment welds and the axial butt welds which terminated at the weld under repair.

3.1.2 Headers

The superheater inlet headers and the evaporator outlet headers were supported from 350mm square 30mm thick pads fillet welded to the boiler shell. It was not practicable to heat the shell via heaters laid over the pads because the interface between the pad and the shell would result in overheating of the pad and the generation of excessive thermal stress in the shell. Furthermore, in the hotter part of the PWHT the attachment would not be able to support the header loads. Thus the headers were disconnected from the pads and re-supported from cold steelwork around the outside of the vessel using spring supports. The centres of the pads were machined away leaving a narrow peripheral band and the fillet weld attaching it to the shell. Heaters were then attached directly to the shell and during reinstatement a replacement pad was welded to the peripheral band.

3.1.3 Drum Risers

The boiler drum riser pipework from the HP boiler and economiser headers were very close to the boiler shell in the areas of the weld repairs. These had to be removed to allow access to the weld area for the machining, welding and inspection operators.

3.1.4 Superheater inlet pipework

The external superheater inlet pipework needed to be removed to allow free access around the shell to install heat treatment equipment and to provide a better working platform height for the welders.

3.2 Internal Mechanical Work

3.2.1 By-pass dampers and associated pipework

Initially it was proposed that the one half of the damper unit and three of the 360mm diameter 14m high by-pass tubes needed to be removed to provide access for internal insulation of the shell. A lifting frame was designed and installed in the top of the boiler and was fitted to existing lugs in the inlet cone. Half of the damper mechanism was removed to expose the top of the by-pass tubes. It became obvious at this point that removal of the first 3 by-pass tubes would not gain enough access and the scope was increased to the first 5 tubes. These were lifted up 600mm and a set of lugs welded on. Lifting tackle was attached to the lugs and the top section cut off and transported out for disposal. This operation was repeated until all 14m of the tube had been removed. During this operation it was discovered that gas baffle plates located between the by-pass tubes and the tube banks were connected to the damper support beams above the tube bank and the tube bank support beams below and would have to removed to allow the tube banks to be lifted. Thus all tubes and the baffle plates were removed. The baffle plates were numbered and stored outside the boiler for re-use and the tubes and the redundant by-pass damper components, were cut up and disposed via an approved route to a Radiological disposal facility at Drigg.

3.2.2 Lifting of superheater tube banks

The superheater tube bank is supported from below by beams which, in turn, are supported by brackets welded to the inside of the shell. During PWHT these brackets would: a) become too hot to support the imposed loads, and b) require direct heating to limit temperature differentials with the shell. Therefore it was proposed to re-support the superheater (100 tonne) within the boiler and disconnect the main beam from the brackets. It was recognised that the boiler internals would be heated during the PWHT operation, accordingly a decision was made to limit the internal temperature to 360 oC and this was used as the design operating temperature of the frame. The resulting frame was made as light as possible by the use of grade 50D steel but even so was a extremely large and heavy structure, Fig. 3. Each component had to be man handled into the boiler, which precluded the use of long fabrication sections, which resulted in a large number of bolted joints.

The lifting frame comprised a main frame slung from brackets bolted to the boiler inlet cone and an intermediate frame slung from the main frame and connected to each individual element by sling rods and clevises. All of the sling rod and clevis arrangements were standard Mitsui Babcock components used in conventional boiler sling decks. The clevices were oriented such that on the main frame they allowed for radial expansion and on the elements they allowed for linear expansion from the fixed thermal sleeves. Each frame and the associated slings were assembled and proof load tested outside the boiler before being installed.

After the frame was installed the load from each element was taken up by tightening the nut on the top of each sling on the elements. In this way the load was taken up in a gradual manner. Each element was only raised by 2-3mm. In order to prove the integrity of the overall arrangement a 10% overload test was carried out by filling the superheater tubes with water. On successful completion of the overload test the tube bank was lifted up to its final position using hollow ram hydraulic jacks on the 4 main sling rods between the main and intermediate frames. The final position was established by considering the relative upward displacement of the support point due to expansion of the shell during PWHT and the downward expansion of the tube bank and supports under internal heating. The criteria being to ensure that the support brackets did not become loaded at any time during PWHT. The superheater auxiliary beams were then disconnected from the main beams and clearance was provided to the auxiliary beam shell brackets and to the superheater main beams by cutting small sections from each end of the auxiliary beams. The superheater main beams were then supported and lifted from an additional frame attached to the top of the main lifting frame.

3.3 Main gas duct
It was considered that disconnection of the main gas duct between the top of each boiler shell and the reactor pressure vessel was appropriate to prevent any loads being applied to the reactor nozzles caused by the expansion of the boiler during the heat treatment. Even though the loads on the pressure vessel were predicted to be low it was judged that the consequences of any uncertainty or a fault condition could have a major effect on the reactor pressure vessel integrity case. By adopting disconnection it could be guaranteed that no additional loads would be transferred onto the reactor pressure vessel.

The joint to be disconnected was on the outlet side of the main gas valve from the reactor pressure vessel, Fig. 1. This consisted of a single flanged connection with a make up ring between the two flanges with a welded seal. A restraint cage was constructed around the joint and the reactor side of the joint was fixed to the lower restraint frame with strain gauged bolts and the boiler side of the joint was fixed to the upper frame in a similar manner. Three-dimensional position-monitoring transducers were attached to both sides of the joint. Four of the flanged studs were also replaced with strain-gauged bolts. This would allow the loads and movements to be accurately recorded when the joint was broken. A tent was built around the joint to contain any radiological contamination that may have been present when the joint was broken. The seal welds were removed by chipping which, although a technique little used today, was judged to be the most suitable as it would produce the least amount of air borne debris. After the joint had been dismantled the maintenance spool piece was removed and the open ends of the duct fitted with temporary blanks, Fig. 4. Only minor movements and loads were recorded confirming that there was very little in-built stress in the system.

The reactor side of the joint was locked for the duration of the repair until the reconnection phase and the boiler side of the ducting was released and allowed to float on the constant load supports with the expansion of the boiler under PWHT.

3.4 Machining of weld excavations
3.4.1 Weld Preparations
Stage 1. Investigative activities.

Prior to undertaking the weld repair excavation samples for metallurgical examination were taken in the vicinity of the defects and beyond their circumferential tips. These samples showed that grain boundary cavitation and micro-cracking was present in the weld HAZ beyond the major crack tips both in the through wall and circumferential directions. In the latter instance this damage extended for a considerable distance and over almost the entire circumference of weld 6/7 on boiler 2C. It was this discovery of the circumferential extent of cavitation in the HAZ that lead to the decision to undertake fully circumferential repairs. Remote from the defects the it was determined that in the through wall direction the damage diminished with increasing depth from the external surface and terminated at ≈15mm deep. The weld repair excavation depth was set at 25mm, with the intention of removing the majority of cavitated material in a single operation. Nevertheless cavitation in the HAZ of the remaining ligament beneath the weld preparation could be prejudicial to the success of the repair welding and metallographic replication of the HAZ on the floor of the weld preparation on all three boilers was undertaken, as described in Stage 3 below.

The HAZ on both sides of the weld were replicated over a circumferential length of 35-40 mm at specific locations around the circumference of each boiler as detailed below:

1. Intervals of at most 500 mm, but frequently smaller, in areas of the excavation corresponding to the major defects and over the full length of the blended transition zones between regions of differing depths.
2. Intervals of 1000 mm along the remaining circumferential extent of the excavation unless damage was observed, in which case the interval was locally reduced.

Acceptance criteria for any damage observed were prescribed as follows:

1. Cracking or micro-cracking (defined as cracks >3 grain diameters in length) were not acceptable unless the latter was isolated and confined within a 1 mm diameter area.
2. Cavitation or decohered grain boundaries (<3 grain diameters in length) were not acceptable in fine grained HAZ (defined as <50 μm grain size). This was on account of the concern that the underlying coarser grained HAZ could be more extensively damaged as it has a greater propensity for stress relief cracking.
3. Localised, circumferentially aligned cavitation and/or decohered boundaries in coarse grained HAZ was deemed to be acceptable providing a 'C' parameter of 0.1 was not exceeded. This 'C' parameter was defined as the summated length of cavitated and/or decohered grain boundaries divided by the total length of coarse grained HAZ area interrogated (2).

Where the above acceptance criteria could not be met, the excavation was deepened by machining and metallographic replication repeated. In the event, all of the criteria were satisfied on all three boilers and circumferentially aligned cavitation and/or decohered grain boundaries were reduced to a 'C' parameter of very much less than 0.1

A number of surface breaking defects revealed by MPI of the floor of the weld repair preparation were removed by hand grinding as were ultrasonically detected buried defects which lay just beneath the floor. These were metallographically replicated and found to be welding induced defects such as entrapped slag and minor hydrogen cracking. No associated stress relief damage was in evidence.

Stage 2 Machining

Although the scope of the weld repair increased from the approximate length of the defects to fully circumferential the machine originally developed to excavate the weld preparations was used although additional machines were acquired to cope with the increased extent. The machine was essentially a hydraulically powered end mill that was moved around the boiler on a curved track. The track being mounted on studs welded to the boiler shell. The machine had the capability of removing the standard excavation in one pass by using purpose made cutters, Fig. 5, which had been manufactured, to the shape of the weld profile. Areas where either deeper or wider excavations were required were machined in a number of passes using horizontal and vertical slide attachments.

Extensive trials of the machines were undertaken by the manufacturer, during which time operators were trained to carry out the work on site.

3.4.2 Weld 5/6 on boiler 2C

This work was undertaken in 5 distinct stages using a modified version of that used for the weld preparation and tested as discussed in the previous section..

Stage 1 Accurately determining depth of defect.

To determine the exact depth of the defect 3 local excavations were made, one at the deepest location, one at the crack tip and one at a shallow location. The maximum depth of excavation was set at 11mm, which was the depth predicted by ultrasonic testing (UT). The material was removed in 2mm increments using a flapper wheel with MPI of the base of the excavation of each increment and replication when there was no MPI response. The defect was almost completely removed at 11mm depth.

Stage 2 Removal of material samples

Samples for mechanical testing and metallurgical examination were removed from within the excavation profile by electro discharge machining (EDM), Fig. 6.

Stage 3 Machining datum face and full survey on boiler surface.

The boiler in the vicinity of the 5/6 weld was not truly cylindrical, it was only slightly out of round but was distorted by the torniquet effect of the original welding operation resulting in a 5mm reduction in radius at the weld centreline. There was also a ~2mm mismatch between the adjoining plates. The finite element analysis of the proposed excavation had shown that there was only 1mm difference between the minimum acceptable ligament thickness and that which would remain after defect removal. Accordingly it was necessary to orient the base of the excavation relative to the inner surface of the shell which involved manipulating arbitrary external data, depth measurements and ultrasonic thickness measurement

Stage 4 Machine base of excavation at final width and depth

The results from stage 3 were utilised to determine the way forward which was to machine to a set depth and locally blend out the remaining defect and the remains of the stage 1 excavation dimples. This was not taken in isolation, as it was imperative that the final surface profile gave good coverage for NDT by ultrasonics. The excavation was not machined to final depth but stopped 1.5mm short. The final 1.5mm was removed by hand polishing to remove any residual stress in the surface of the material left from the machining process.

Stage 5 Machine sloping sides of excavation.

After the base of the excavation had been formed a full NDT examination of the excavation was undertaken before the sloping sides were formed. The sloping sides were machined again to within 1.5 mm of the final depth and the remaining material removed by hand dressing. A full thickness and surface profile was taken of the finished excavation to show that the correct amount of material had been removed by hand dressing and also to record the final results for the case history.

As this area of the work was so critical to the overall weld success of the repair project all operations were first undertaken on a shell mock up to prove the capability, each operation being personally witnessed by the ITPIA.

3.5 Repair Welding

The site welding activities started with a rigorously defined weld procedure and highly skilled welders following the weld procedure development (2) and welder training activities (3) described in other papers. The two layer refinement technique is not particularly novel but it is unlikely that it has ever been so rigorously controlled on such a large repair.

The welding electrodes were manufactured by MBEL in a limited number of batches which had been individually tested in accordance with ASME requirements. They were supplied in vacuum packs and a dedicated environment controlled electrode store was established close to the workface. Procedures were put in place to check each pack for integrity prior to issue to welders, limit issue to one pack at a time to a welder, dispose of all unused electrodes and stubs at the end of a shift and record the welder ID, weld deposit location and pack ID for all electrodes used. Rebaking of unused electrodes at site was not permitted (unused electrodes were identified and stored and if rebaking had been required they would have been returned to the manufacturing location and baked, and vacuum packed and tested as if new batches).

The weld length was divided into 24 zones, the objective being that the maximum number of welders could work simultaneously. Successive shifts would carry on from the previous shift yet overall complete the weld one layer, at a time, Figs. 7 and 9, and have only one welder working on a given section of weld to aid traceability if defects should occur. Some compromises had to be made in order to maintain reasonable progress but any such events had to be approved in advance and were fully recorded.

The deposition parameters of the first two and, to a lesser extent the 3rd and 4th layers, were critical and closely monitored. During this phase each welder was accompanied by a monitor who calculated heat input from the PAMS monitor output and confirmed the run out ratio and the bead overlaps for every weld run. Additionally, at the end of each shift, the welding parameter data were entered onto a PC which confirmed that the manual calculations were correct.

Beyond the deposition layer the detailed recording of weld parameters was not considered necessary, other than random checks and no checks were made beyond the fourth layer. Nevertheless, by normal welding standards rigorous controls were maintained on electrode storage and issue, deposition pattern and welding technique.

The welding operation was completely successful and is a testament to the procedure development, expertise of the welders and the welder training programme. Of approximately 10,000 individual weld deposits (weld rods) in the first two layers only <100 deposits were removed due to out of tolerance deposition conditions, and the majority of these occurred in one area in the first repair due to a misinterpretation of the specification which was subsequently clarified. Only 3 defects were detected in the repair weld deposit by NDT, the largest which was 60mm long, intermittent, < 3 mm high, which is remarkable by any standard for MMA welding.

3.6 Preparations for PWHT
3.6.1 Power supplies
The full load capacity required for the proposed Heat Treatment was estimated at 5MW allowing pre-heat and welding on one boiler shell in parallel with PWHT on another.

It was decided to utilise the 11kV supply from the 2C main gas circulator in a split-ring main system connected to 5 manually switched 11kV/415v Transformers, two 1.5Mva and three 1Mva. The twelve 415v outlets were connected via twelve changeover switches to enable supplies to be switched as required to the two boiler houses, dependant upon the operations being carried out. Twenty four 400mm 415v cables were cleated every metre and affixed to two specially designed 50 metre high scaffold towers, Fig. 8, and terminated onto a scaffold platform suspended from the Boiler House roof beams. The twelve distribution panels were then connected to a series of 77 415/110v transformers which provided the supplies to allow the Heat Treatment processes. The installed system included 3.4km of 400mm armoured cable, 4km of ladder racks and over 3,500 cable clips with a total weight of 110 tonnes. All the cabling had to be hand pulled into the building, including the 111 off heat treatment 415/110v transformers which had to be manually placed at the top of each boiler house. Over 88km of heat treatment 6mm cable was installed. The system performed as designed, not one trip or loss of supply was experienced.

It was originally envisaged that the entire electrical equipment would be located in the boiler house on the floor of the Top Duct. Prior to installation the weights were assessed and found to exceed the maximum safe loading for the floor. It was necessary to reinforce the boiler house roof beams and mount part of the electrical equipment on a suspended platform. The 415v/110v transformer, monitoring and control equipment were then positioned on the reinforced top duct floor in a specific layout to suit the heat treatment process.

Because of the specific repair programme and the necessity to increase the number of 415/110v transformers during the Post Weld Heat Treatment, it was necessary to man handle 40 transformers from one boiler house to the other and then to reverse the situation, for the next Post Weld Process.

3.6.2 Internal insulation
One of the early considerations had been to remove all of the superheater elements through the gas circulator opening in the bottom of the boiler in the same manner that the boiler had been originally constructed. This would have removed the requirement to de-load the superheater support brackets and incidentally given free access to the inner surface of the boiler for fitting thermal insulation. This option although practical would have required a significantly longer programme time due to the removal of the gas circulator and the associated problems of the storage of large quantities of radiologically contaminated

components. Instead a thorough examination of the construction of the boiler and the access that could be obtained lead to the conclusion that internal insulation could be installed with the tubes in place.

Detailed proposals for internal insulation were made during the development phase (2). Installation of the superwool matting on accessible areas of the shell proceeded largely as intended. There were some difficulties assembling and inserting the stainless steel / microtherm panels into the narrow gap between the tube banks and the shell, largely because it had not been possible to carry out detailed surveys or trial assemblies prior to starting the work. There were some installation problems which were difficult to identify because of the narrow curved gap between the tube bank and the shell.

The main technique used to identify the nature of problems was the use of miniature TV cameras with a variety of camera/light arrangements and delivery systems. It became standard practise to carry out a survey of all locations prior to attempting to install insulation. The majority of installation difficulties were due to individual tubes being slightly out of position, maybe only 10-20mm which is a perfectly acceptable tolerance for boiler construction but could significantly alter clearances for insulation. Fortunately the arrangements made to lift the tube banks removed the rigid connections between platens allowing them to be individually raised and lowered or moved laterally.

It must be borne in mind that all of the work was carried out in C3 radiological conditions with extremely limited access and working room, as shown in Fig 10.

On completion of the PWHT examination of the internal insulation showed that it had remained intact and that the various retention and spring loading systems had operated correctly. With only one exception which was resolved by inspection and manipulation all of the insulation panels were easily removed.

3.6.3 Boiler cooling system
A boiler cooling system, was installed to counter any adverse temperature within the boiler during heat treatment. Cooling air entered in through the boiler 85' level mandoor and exiting the circuit through the 145' level mandoor, Fig. 1. An additional inlet point at the 175' mandoor provided tempering air to cool the hot air from the boiler before it entered the filters. Power was provided by utilising 3 of the standard extraction units, which were used to maintain a negative pressure in the boiler during entries to prevent the spread of radiological contamination. Dampers were fitted at the air inlets to allow the flows to be balanced. The extracted air was ducted to the boiler house roof where it was exhausted to atmosphere. The absolute (HEPA) filters on the extraction units prevented any radiological contaminated particles from being discharged from the boiler. One fan was in constant operation throughout the heat treatment cycle to maintain a negative pressure inside the boiler with the inlet damper closed. In the event that the temperature inside the boiler started to rise beyond $290^{\circ}C$ then the inlet damper would have been opened to induce a cooling flow through the boiler. If this had been insufficient the second or third extraction units could have been started.

3.6.4 Instrumentation
A suite of instrumentation was fitted to the boiler and ducting to measure transient conditions during the heat treatment. These were all centrally logged during the heat treatment cycle to show how the plant had moved and also to prove that there had been no adverse loading on any of the components. The main instrumentation systems are detailed below.

High temperature strain gauges were fitted to the superheater lifting frame to show that the load had not been transmitted back down onto the shell brackets during the heat treatment cycle.

High temperature strain gauges were fitted to the boiler inlet nozzle to show that no adverse loadings had been applied to this area by the expansion of the boiler.

Position monitoring was fitted to the free end of the gas duct and to the boiler inlet nozzle to monitor movement and expansion. See Fig. 4.

Thermocouples were fitted at various locations within the boiler to measure the temperature of the air and the lifting frame.

These results, which were overall consistent with expectation, used to justify the integrity of the plant following the repair

3.7 Heat treatment

The statistics elsewhere in this paper give some idea of the magnitude of the equipment/cabling installation. For a single PWHT there were 77 415v transformers over 40 temperature recorders, ~ 2000 heater elements in heaters, each with its own power supply, control thermocouple and monitoring thermocouple, controller and high current relay, not including spares, Figs. 11 and 12. This equipment was installed and verified in accordance with detailed method statements and quality plans. As a final check of system integrity individual heaters were operated briefly to confirm that they and their monitoring systems were correctly connected. This operation detected only a negligible number of errors or poor connections.

A team of heat treatment technicians installed all of the equipment from the 415v transformers onwards, checked it was correct and then conducted the PWHT operation. For the PWHT they were divided in to a number of sub teams each of which was responsible for one or more bands of heaters. Within each team members were allocated specific roles for which they received additional training: This included:

- monitoring temperatures against the intent and identifying any untoward behaviour within the system

- making adjustments to controls both at the predefined rate change points or in response to requests from the monitors

- replacement or rectification of defective equipment

Prior to each PWHT there was a team talk to confirm that all arrangements were in place, ranging from a review of the weather forecast and grid reliability to confirmation of necessary documentation and project approval to commence PWHT.

Following the welding and inspection activities and replacement of the temporary preheat heater elements the band adjacent to the weld was at 150^0 C-200^0C but the outer edges of the heated zone were at ambient temperature. The first operation of the PWHT cycle was to raise all of the heated bands to 200^0C to establish a constant starting temperature for the heating cycle.

From 200^0C the heating rate at the weld was 50 C/hr to 350^0C. The purpose of the 'slow' rate was to allow time for any problems to be resolved without giving rise to any out of specification situation. Even though the system had been rigorously tested this was the first time that it had operated at significant power levels. At 350^0C the heating rate was increased to 98^0C/hr until the weld reached 650^0C and then reduced to 50^0C/hr until the soak temperature $(650 \pm 10^0$C) was reached. The reduction in heating rate towards the end of the heating cycle was a compromise between passing through the stress relief cracking susceptibility temperature range as fast as possible and not overshooting the final soak temperature. The temperature tolerance band was $\pm10^0$C at soak, compared with $\pm20^0$C during heating, and the reduced rate during the final heating phase allowed the necessary fine tuning to be able to comply with the soak temperature limits from the outset. The soak period started when the last monitoring thermocouple within the hot zone (the band of heaters on the weld centreline and the two adjacent bands) attained 665^0C.

The desired final temperature profile during heating and a soak and the principles of the control and monitoring activities have been described in another paper (2). This section describes how these objectives were attained. The parameter which dictated the temperature at any position within the heated zone was the mean temperature at the weld. Operators were provided with graphs and tables from which they could determine the desired mean temperature of any other band for the prevailing weld temperature. The current weld average temperature was determined by the PC system and shown on a large digital display unit visible to all operators from their workstations. Fig. 13 shows the axial temperature profile during the heating phase.

Independently the PC system provide the lead team with a summary of the status of the heat treatment both as an overview of band temperature compared to the calculated ideal value and as a print out of every monitoring thermocouple value. The significant benefit of the PC system is that it clearly flagged any data as warning (within 5^0C of the limiting values), or alarm (outside the limiting values) status. Thus, to monitor the progress of the PWHT a glance at the row summary sheet would establish the presence or otherwise of warning or alarm flags. It a flag was present then the detailed print out for that row would identify the relevant (thermocouples)(s). The system meant that the lead team could remain detached from the detailed operations and concentrate on trends and incidents as they evolved.

Although the heating part of the PWHT cycle had been subject to rigorous development the cooling phase was less well quantified. The main reasons for this was that as residual stresses had been eliminated and the weldment tempered there was no further risk of stress relief cracking and thus cooling rates were not important. It is acknowledged that metallurgical changes will occur during slow cooling, but they were second order effects and would be quantified elsewhere (4). In order to prevent excessive thermal stresses it was intended to reverse the development of the axial temperature profile during the heating phase and the overall cooling rate would be limited by the area with the slowest natural cooling rate.

Initially a programmed cooling rate of 50^0C /hr was set, but it became obvious that this could not be attained in some areas with the result that excessive thermal gradients would be generated. By trial and error revised cooling rates which did not give rise to excessive gradients were determined. However, the cooling rate was very slow, of the order of 7^0C/hr. In order to increase the cooling rate the cooling cycle was placed under manual control. The set point temperature (the temperature the controller will attempt to maintain) was dropped for all rows by an amount which would not give rise to unacceptable thermal gradients if local areas cooled very rapidly. As cooling occurred this process was repeated, the objective being to maximise temperature differences thereby maximising the cooling rate. The decisions on the magnitude and frequency of the reductions in set point temperature were made by the lead team on the basis of the PC data presentation and maintaining a 'running plot' of the axial temperature profile.

Originally it had been intended to stop controlling the heating operation once the temperature reached 430^0C, as permitted by ASME. However, when this was first attempted it was clear that large thermal gradients could occur and control was reinstated down to 300^0C, well within the normal plant operating temperature over the whole heated band and down to 200^0C in the vicinity of the weld. As the temperature reduced to allowable temperature gradients were expanded on the basis that the material yield strength was increasing and this helped maintain the cooling rate. Even so the maximum cooling rate that could be attained was 17^0C /hr. A typical weld centreline time - temperature profile through the PWHT operation is shown in Fig. 14 and the variation in axial temperature profile with time during cooling is shown in Fig. 15.

The internal air and metal temperatures did not exceed 290^0C and thus the venting and cooling system was not required although the temperature trends indicated that had the heating/soak time been longer it would have been needed. However, it was of benefit during the cooling phase and on the basis of experience during the first PWHT it was operated at maximum capacity from the start of soak for the second and third boilers.

On completion of each PWHT all of the temperature data was analysed and assessed, the PC system was invaluable in the analysis activity with some 30,000 data values to be assessed. All alarm conditions were compared with hypothetical temperature gradients which had been deemed acceptable by structural analysis prior to the PWHT operation. Local temperature gradients within rows and between rows were also determined and evaluated. None of the observed gradients required additional structural integrity analysis.

Overall the PWHT operations for all three boilers were completed successfully. On the first heat treatment a plant problem caused the heat treatment to be paused at 350^0C for 18 hours, but was then completed as intended, During the period from the start of heating to the end of soak for all 3 heat treatments only 51 of the 3000 monitoring thermocouples attained alarm status. The majority of these events were brief transients caused by minor equipment or control problems which were rapidly rectified. The remainder were due to a deliberate intent to run a section of heaters slightly above the upper limit in order to assist adjacent heaters maintain more important temperature differentials. None of the departures from the specification gave rise to temperature gradients greater than had been assessed, and deemed acceptable prior to the PWHT operation.

4.   REINSTATEMENT

4.1 Re instatement controls
The re instatement was conducted with the ITPIA carefully checking each item of the works. As each step of the work progressed the ITPIA signed a dedicated Quality Plan confirming an inspector had witnessed the work completed to their satisfaction.

Final checks within the Boiler Shells, throughout the boiler houses were carried out by three levels of inspection, a MBSEL Engineer, a Magnox Engineer and a member of the ITPIA, each signing the check sheet which was part of the Quality Plan, each check being closed out only when the check sheet was completely signed.

4.2 Gas duct reconnection
The gas duct reconnection proved to be a straightforward reversal of the disconnection process. The ducts had moved very little during the heat treatment process and re-alignment did not present any difficulty. The main problem experienced was the replacement of the original flange bolting. The original studs were manufactured from a Babcock and Wilcox specification material BW9 that had a very high strength. In order to manufacture the new stud material had to be procured with a similar chemical composition and heat treat the blanks to obtain the mechanical properties, in essence re-qualifying the material to BW9.

4.3 Reloading of superheater tube bank
This operation was straightforward and no undue problems were encountered. The hydraulic jacks were not used to lower the tube banks as this would have not provided the degree of control that was required and to minimise any shock loading onto the shell brackets or support beams, screw turn buckle arrangements were used.

Initially the tube bank support beams were reinstated and the cut sections of beam carried out during dismantling were welded back in position. The welds tested volumetrically. Each element was lowered individually 2-3mm at a time sequentially across the whole bank, this prevented fouls between the extended surfaces of adjacent elements. Feeler gauge checks were used to ensure that the elements had been fully seated onto the support beams before the lifting frame was removed.

The removed sections of external superheater pipework and headers were replaced and the drum riser pipework with new hanger supports were reinstated. No difficulties were encountered further demonstrating that the boiler shell had returned correctly to its original position following heat treatment.

A hydraulic overpressure test of the superheater tubes was undertaken after all of the pipework had been replaced which proved that the plant had not suffered any adverse affects from the lifting and heat treatment operations.

4.4 Gas by-pass blanking and baffle reinstatement
With the originally designed gas by-pass system being operationally redundant, theoretically all that was required was to blank off the openings between the tube banks to ensure that the gas passed over the heating surfaces. However, this would have changed the shape of the boiler internals and could have altered the gas flow path resulting in possible vibration problems. There were two options available. Either to undertake a complex analysis of the flow and dynamics of the boiler to prove that the new arrangement was acceptable or to install a blanking system that would mimic the original design arrangement. The latter option was taken and justified on time and cost.

During the reinstatement work the components were designed to be fabricated outside the boilers and to be bolted together in situ to reduce the time spent inside the boiler and hence to reduce the total radiological dose to the operatives in line with the ALARP principal. Welding was avoided wherever possible inside the boilers to reduce the chances of stray arc strikes on the tubes or risk of fire. The material chosen for the new boiler internal components was a boiler plate with a controlled silicon content as this has proven to offer resistance to oxide growth which is a problem that has affected a number of Magnox boilers.

The bolts used were the modern equivalent of the original construction materials and rolled threads were specified for better ductility.

5.   RECOMMISSIONING AND TEST

Detailed recommissioning procedures were prepared and agreed with Magnox safety assessors and the NII.

The recommissioning process was managed by two important committees.

(i) The Reactor 2 Commissioning Committee comprising of Station Personnel, Contractors and Safety Assessors, chaired by the Station Manager.

The committee's key responsibilities were to ensure that all aspects of Nuclear Safety requirements for the Project had been satisfied and to ensure that all other issues in bringing the Reactor back to power had been properly addressed in accordance with Site License conditions.

(ii) The Reactor 2 Test and Commissioning Panel whose membership included Station Project and Contract Staff chaired by the Site Project Manager.

The committee's prime responsibilities were to ensure that all work carried out was in compliance with agreed procedures and to manage the test and commissioning of the repaired boilers back to full operational duty. The panel formally reported to the Reactor 2 Commissioning Committee.

Two significant parts of the recommissioning process were:

(i) Hydraulic over pressure test of each boiler's steam and water circuits pipework, drums and headers.

(ii) Low pressure air test and subsequent external leak checks of the CO_2 coolant gas path of each boiler

All the tests were completed on each repaired boiler with very satisfactory results.

All the remaining checks and tests carried out on repaired boilers, which included strain gauge monitoring, vibration and gas duct movement and hysteresis checks similarly gave very satisfactory results.

The repaired boilers were returned to service without any operational difficulty all installed instrumentation gave expected readings and no abnormalities were noted.

6. MAJOR STATISTICS

- million man hours worked.

- personnel site team working 24 hours (over 750 workers involved)

- Over 1,000 Safety Case documents produced.

- Over 2,000 Work Control Cards issued and risk assessed.

- Over 6,000 Boiler Entries made under C3 conditions.

- Over 6,000 electrical connections to the heat treatment equipment and over 5,600 heat treatment heater pads.

- Over 180 Transformers used to provide lighting, welding, heat treatment and general supplies.

- Over 12,200 tons of scaffolding materials.

- Over 88km of heat treatment cabling (6mm).

- Over 3.4km of 415v cables used to provide 4.5Mva to the boiler houses (400mm)

7. CONCLUSION

The boiler repair project at Sizewell was a major success both in the achievements and also in the close working relationship that was developed between the Magnox Electric and Mitsui Babcock. The principal successes of the Project were:

- No lost time accidents Gold ROSPA awarded.

- Weld repairs implemented to a very high Quality Standard.

- Meeting site programme requirements with an excellent team spirit.

- Case history completed and accepted within 7 days of last entry into the boilers.

- Delivery of 3 successfully repaired boilers released to service.

8. ACKNOWLEDGEMENT

In addition to the numerous personnel from Magnox, Mitsui Babcock and the other organisations mentioned in this paper the following contributions are acknowledged:

- Reekie machining for weld 6/7 repair and weld 5/6 blending excavations.

- The Didcot Heat Treatment team led by D Hobson with P Thompson and S Smith for the operation of the heat treatment.

- The Fincham Insulation team led by P Murphy for the installation of the internal insulation.

This paper is published with the approval of the Director of Technology and Central Engineering, BNFL Magnox Generation.

9. REFERENCES

1. Jeans P.J., Taylor E.G. and Munro H.G. Paper 5, these proceedings.
2. McDonald E. J., Hunter A.N.R., and Bell W. Paper 6, these proceedings.
3. Tolaini J. Paper 12, these proceedings.
4. Hunter, A.N.R., McDonald E.J., Moskovic R. and Lamb M. Paper 8, these proceedings.

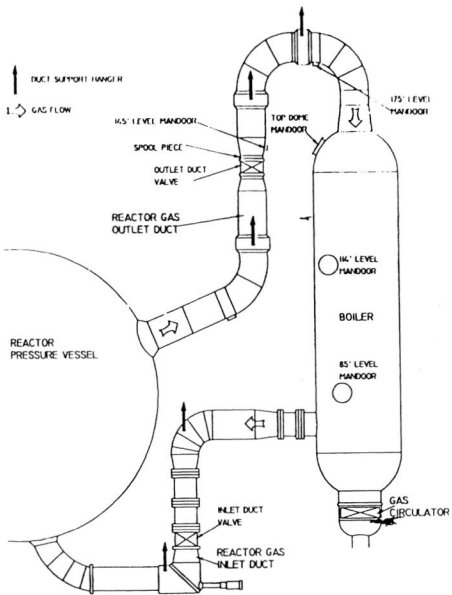

Figure 1: Schematic layout of gas circuits at Sizewell A

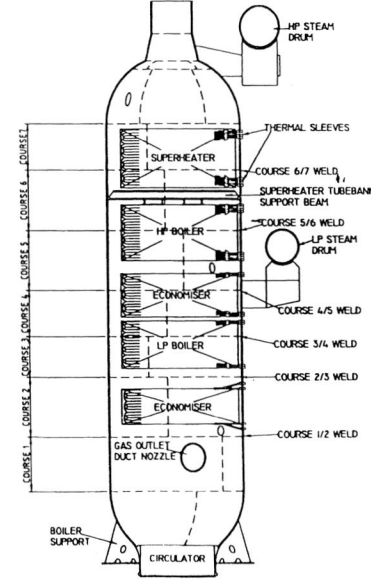

Figure 2: Schematic layout of boiler at Sizewell A

Figure 3 Lifting frame installed in boiler 2C

Figure 4 Disconnected gas duct with spool removed

Figure 5 Machining weld excavation

Figure 6 Removing material samples from weld 5/6

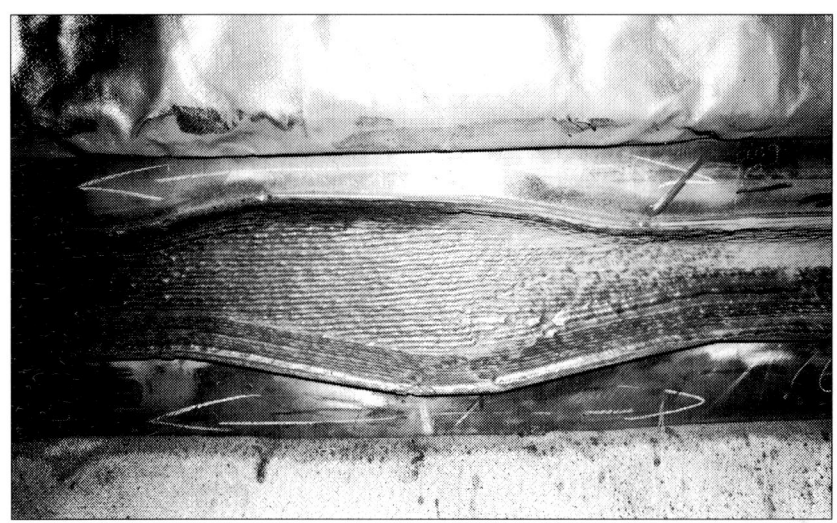

Figure 7 First layer completed on wide excavation and second layer started

Figure 8 Temporary power cables to heat treatment stations

Figure 9 Welding of repair in progress

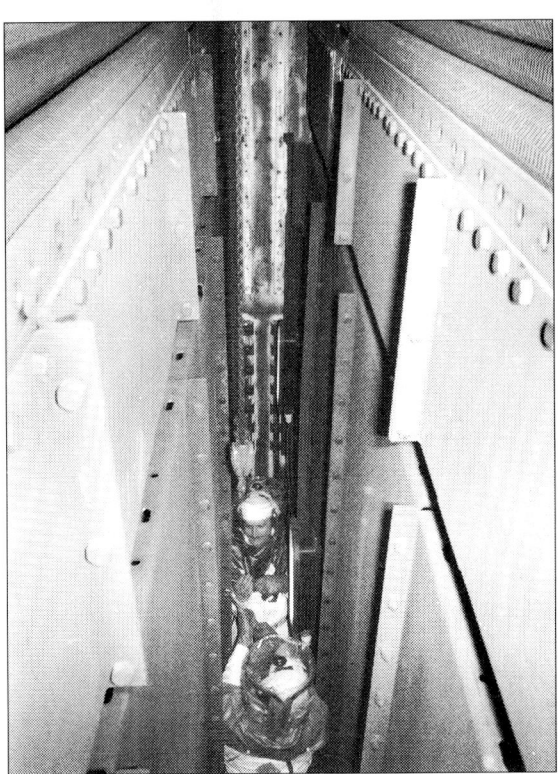

Figure 10 Fitting insulation between the tube bank and the boiler shell

Figure 11 Installation of heater mats to boiler shell

Figure 12 Heat treatment station

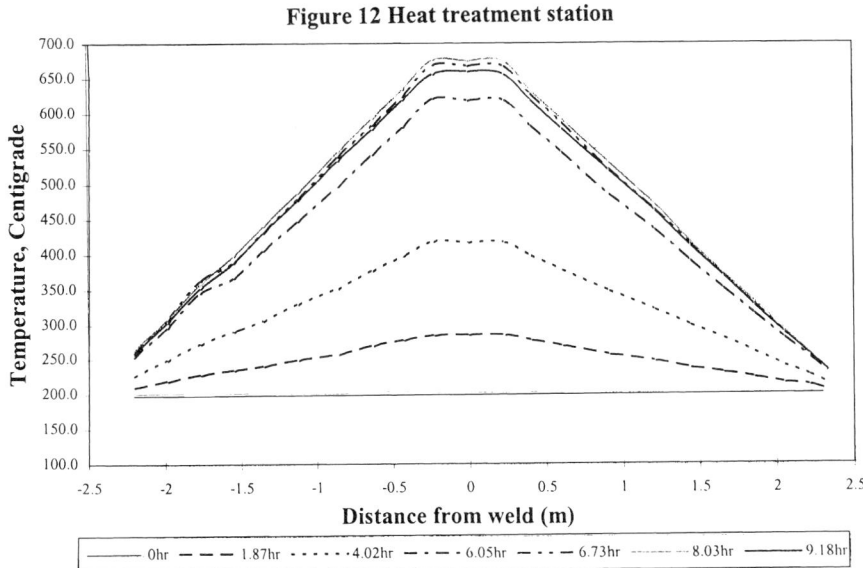

Figure 13 Axial temperature profile; start of PWHT to end of soak.

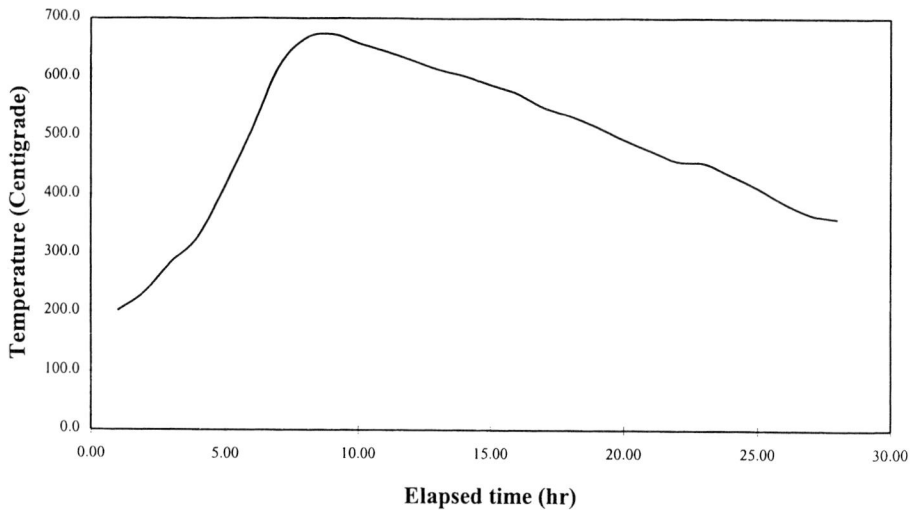

Figure 14 Variation in repair weld temperature during PWHT.

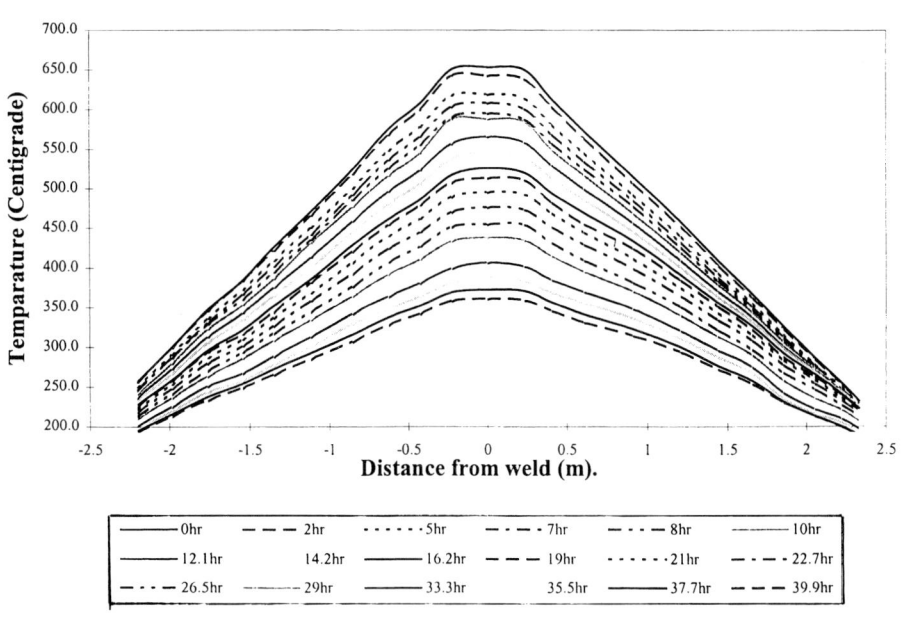

Figure 15 Axial temperature profile during cooling from end of soak.

S690/009/99 *Sizewell A Power Station Boiler Repair*

Authors' Index

.